Preventing Lead Poisoning in Young Children

A *STATEMENT BY THE*
CENTERS FOR DISEASE CONTROL AND PREVENTION
AUGUST 2005

U.S. DEPARTMENT OF HEALTH AND HUMAN SERVICES
Centers for Disease Control and Prevention

Preventing Lead Poisoning in Young Children

A Statement by the Centers for Disease Control and Prevention
August 2005

Centers for Disease Control and Prevention
Julie L. Gerberding, MD, MPH, Director

National Center for Environmental Health
Tom Sinks, PhD, Acting Director

Division of Emergency and Environmental Health Services
Jim Rabb, Acting Director

Lead Poisoning Prevention Branch
Mary Jean Brown, ScD, RN, Chief

U.S. Department of Health and Human Services, Public Health Service

Suggested reference:
Centers for Disease Control and Prevention. Preventing Lead Poisoning in Young Children. Atlanta: CDC; 2005.

Table of Contents

Advisory Committee on Childhood Lead Poisoning Prevention Members, Ex-Officio and Liaison Representatives

CHAIR
Carla Campbell, MD, MS
The Children's Hospital of Philadelphia
Philadelphia, Pennsylvania

EXECUTIVE SECRETARY
Mary Jean Brown, ScD, RN
Chief, Lead Poisoning Prevention
  Branch
National Center for Environmental
  Health
Centers for Disease Control
  and Prevention
Atlanta, Georgia

MEMBERS
William Banner, Jr. MD, PhD
The Children's Hospital at Saint Francis
Tulsa, Oklahoma

Helen J. Binns, MD, MPH*
Children's Memorial Hospital
Chicago, Illinois

Walter S. Handy, Jr., PhD
Cincinnati Health Department
Cincinnati, Ohio

Ing Kang Ho, PhD
University of Mississippi Medical
  Center
Jackson, Mississippi

Jessica Leighton, PhD, MPH
New York City Department of Health
  & Mental Hygiene
New York, New York

Valarie Johnson
Urban Parent to Parent
Rochester, New York

Tracey Lynn, DVM, MPH†
Alaska Department of Health
  and Social Services
Anchorage, Alaska

Sally Odle
SafeHomes, Inc.
Waterbury, Connecticut

George G. Rhoads, MD, MPH
University of Medicine and Dentistry
  of New Jersey
Piscataway, New Jersey

Catherine M. Slota-Varma, MD
Pediatrician
Milwaukee, Wisconsin

Wayne R. Snodgrass, PhD, MD
University of Texas Medical Branch
Galveston, Texas

Kevin U. Stephens, Sr., MD, JD
New Orleans Department of Health
New Orleans, LA

Kimberly M. Thompson, ScD
Harvard School of Public Health
Boston, Massachusetts

_____
*ACCLPP member until May 2004
†ACCLPP member until October 2004

Alan Pate
Battelle Memorial Institute

Joel Schwartz, Ph.D.
Department of Environmental Health
Harvard School of Public Health

David Bellinger, PhD,
Neuroepidemiology Unit Children's
    Hospital
Harvard Medical School

David A. Savitz, PhD
Department of Epidemiology
University of North Carolina School
    of Public Health

Carla Campbell, MD, MS
Division of General Pediatrics
The Children's Hospital of Philadelphia

Patrick J. Parsons, PhD
Wadsworth Center for Laboratories
    and Research
New York State Department of Health

Betsy Lozoff, MD
Center for Human Growth and
    Development
University of Michigan

Kimberly Thompson, ScD
Department of Health Policy
    and Management
Harvard School of Public Health

Birt Harvey, MD
Pediatrician
Palo Alto, California

Preface

This is the fifth revision of *Preventing Lead Poisoning in Young Children* by the Centers for Disease Control and Prevention (CDC). As with the previous statements, the recommendations presented here are based on scientific evidence and practical considerations. This revision accompanies a companion document, *A Review of Evidence of Adverse Health Effects Associated with Blood Lead Levels <10 µg/dL in Children*, developed by Advisory Committee on Lead Poisoning Prevention which reviews the scientific evidence for adverse effects in children at blood lead levels below 10 µg/dL.

The data demonstrating that no "safe" threshold for blood lead levels (BLLs) in young children has been identified highlights the importance of preventing childhood exposures to lead. It confirms the need for a systematic and society wide effort to control or eliminate lead hazards in children's environments before they are exposed. This emphasis on primary prevention, although not entirely new, is highlighted here and is clearly the foremost action supported by the data presented in *A Review of Evidence of Adverse Health Effects Associated with Blood Lead Levels <10 µg/dL in Children*.

Although there is evidence of adverse health effects in children with blood lead levels below 10 µg/dL, CDC has not changed its level of concern, which remains at levels ≥10 µg/dL. We believe it critical to focus available resources where the potential adverse effects remain the greatest. If no threshold level exists for adverse health effects, setting a new BLL of concern somewhere below 10 µg/dL would be based on an arbitrary decision. In addition, the feasibility and effectiveness of individual interventions to further reduce BLLs below 10 µg/dL has not been demonstrated.

CDC is conducting several activities to focus efforts on preventing lead exposures to children. First, beginning in 2003, CDC required state and local health departments receiving funding for lead poisoning prevention activities to develop and implement strategic childhood lead poisoning elimination plans. Second, CDC and its federal partners, the Department of Housing and Urban Development and the Environmental Protection Agency, launched new initiatives to control lead-based paint hazards in the highest risk housing, addressing where successive cases of lead poisoning have been identified. Third, CDC and other federal agencies are developing a systematic and coordinated response to identify and eliminate non-paint sources of exposure (e.g., lead jewelry, food and traditional medicines, and cosmetics).

CDC continues to monitor progress toward the Healthy People 2010 objective of eliminating elevated BLLs in children at the national level through the National Health and Nutritional Examination Survey and at the state and local levels through the blood lead surveillance system. These complementary data provide essential information for the rational distribution of resources to communities with the highest risk for lead exposure.

I wish to thank both current and former members of the Advisory Committee on Childhood Lead Poisoning Prevention and consultants who developed the documents in this statement and acknowledge their contribution to the health of the nation's children. The Committee considered a number of controversial issues, examined the existing data and reviewed the report of the work group. This 2005 statement represents agreement of 12 of the 13 Advisory Committee members serving on the committee as of October 19-20, 2004.

Thomas Sinks, Ph.D.
Acting Director
National Center for Environmental Health/Agency for Toxic Substances and Disease Registry

INTRODUCTION

The U.S. Department of Health and Human Services has established an ambitious goal of eliminating elevated blood lead levels (BLLs) in children by 2010, a qualitatively different goal from earlier goals that focused on reducing the BLL considered toxic by various target amounts.[1] Recent research on lead's health effects at low levels, which suggests societal benefits from preventing even low level lead exposure in childhood, underscores the importance of this public health goal.

This revised statement describes the public health implications of research findings regarding adverse health effects at low BLLs summarized in the accompanying review, and focuses on the Centers for Disease Control and Prevention's blood lead "level of concern." This statement aims to guide public health practice and policy development and review necessary steps to ensure progress toward meeting the 2010 goal.

PREVENTING CHILDHOOD LEAD POISONING IN THE UNITED STATES

The reduction of BLLs in the United States during 1970-1999, primarily because of implementation of federal and state regulations to control lead exposure, was one of the most significant public health successes of the last half of the 20th century.[2] Nonetheless, some populations and geographic areas remain at disproportionately high risk for lead exposure.[3-5] Specific strategies that target screening to high-risk children are essential to identify children with BLLs ≥10 µg/dL. Once identified, children with elevated BLLs should receive follow-up services as recommended in *Managing Elevated Blood Lead Levels Among Young Children*.[6]

However, *preventing* elevated BLLs is the preferred course of action. A compelling body of evidence points to the limited effectiveness of waiting until children's BLLs are elevated before intervening with medical treatments, environmental remediation, or parental education. [7-12] Data indicate that in many cases it takes years to reduce children's BLLs once levels are elevated whether the initial blood lead elevation is very high or moderate.[13-15] The most common high-dose sources of lead exposure for U. S. children are lead-based paint and lead-contaminated house dust and soil. Recent studies have identified methods to reduce common household lead hazards safely.[16] Thus, a multitiered approach that includes secondary prevention through case identification and management of elevated BLLs is needed to eliminate childhood lead poisoning. However, because no level of lead in a child's blood can be specified as safe, primary prevention must serve as the foundation of the effort.

CDC'S BLOOD LEAD LEVEL OF CONCERN

The adverse health effects associated with elevated BLLs have been widely studied and documented. Previously, CDC responded to the accumulated evidence of adverse effects associated with lead exposures by lowering the BLL of concern. Between 1960 and 1990 the blood lead level for individual intervention in children was lowered from 60 µg/dL to 25 µg/dL. In 1991 the CDC recommended lowering the level for individual intervention to 15 µg/dL and implementing community-wide primary lead poisoning prevention activities in areas where many children have BLLs ≥10 µg/dL.[17] Some activities, such as taking an environmental history, educating parents about lead, and conducting follow-up blood lead monitoring were suggested for children with BLLs of ≥10 µg/dL.[7] However, this level, which was originally intended to trigger communitywide prevention activities, has been misinterpreted frequently as a definitive toxicologic threshold.

As the accompanying review of recent studies indicates, additional evidence exits of adverse health effects in children at BLLs <10 µg/dL. The available data are based on a sample of fewer than 200 children whose BLLs were never above 10 µg/dL and questions remain about the size of the effect.

At this time there are valid reasons not to lower the level of concern established in 1991 including the following:

- No effective clinical or public health interventions have been identified that reliably and consistently lower BLLs that already are <10 µg/dL. Nonetheless, the sources of lead exposure and the population-based interventions that can be expected to reduce lead exposure are similar in children with BLLs <10 µg/dL and ≥10 µg/dL,[18] so preventive lead hazard control measures need not be deferred pending further research findings or consensus.

- No one threshold for adverse effects has been demonstrated. Thus the process for establishing a lower level of concern would be arbitrary and no particular BLL cutoff can be defended on the basis of the existing data. In addition, establishing a lower level of concern may provide a false sense of safety about the well being of children whose BLLs are below the threshold.

- The adverse health effects associated with elevated BLLs are subtle. Individual variation in response to exposure and other influences on developmental status, make isolating the effect of lead or predicting the overall magnitude of potential adverse health effects exceedingly difficult.

- Establishing a level of concern substantially <10 µg/dL probably would be accompanied by a sharp increase in misclassification of children as having an elevated BLL. The uncertainty associated with laboratory testing is too great to ensure that a single blood lead test reliably classifies individual children at levels <10 µg/dL. This misclassification could confuse both parents and clinicians and expenditure of resources on testing that does not aid decision making.

- Efforts to identify and provide services to children with BLLs <10 µg/dL may deflect needed resources from children with higher BLLs who are likely to benefit most from individualized interventions.

- Efforts to eliminate lead exposures through primary prevention have the greatest potential for success. Reducing exposures will benefit all children, regardless of their current BLL.

RESPONDING TO DATA ON ADVERSE HEALTH EFFECTS AT BLOOD LEAD LEVELS <10 µg/dL FROM A PUBLIC HEALTH PERSPECTIVE

Since 1991, CDC has emphasized the need to make primary prevention of lead poisoning, through interventions that control or eliminate lead hazards before children are exposed, a high priority for health, housing, and environmental agencies at the state, local, and federal levels.[18-20] Federal and state policies and programs, largely as the result of Title X of the 1992 Housing and Community Development Act (Public Law 102-550), increasingly have focused on the need for primary prevention using strategies known to effectively reduce residential lead hazards.[21] Research findings also indicate that primary prevention would be expected to benefit all children at high risk because communities with the largest percentages of children with BLLs ≥20 µg/dL also have the largest percentage of children with BLLs that are lower but still above the national average of approximately 2 µg/dL.[18] These data underscore the importance of targeting efforts to communities where risk for exposure is highest and provide a strong rationale for primary prevention efforts. The strategies described below will effectively direct efforts to achieve the Healthy People 2010 objective to eliminate lead poisoning in young children and can be expected to reduce lead exposure for all children.[1]

Primary Prevention

CDC's Advisory Committee on Childhood Lead Poisoning Prevention recently issued updated recommendations calling for the nation to focus on primary prevention of childhood lead poisoning.[22] Because the 2010 health objective of eliminating childhood lead poisoning can be achieved only through primary prevention,[22] this document provides important guidance to state and local agencies regarding the implementation of primary prevention activities. Given that the most important measure of a successful primary prevention strategy is elimination of lead exposure sources for young children, we focus here on the two main exposure sources for children in the United States: lead in housing and non-essential uses of lead in other products.

Lead in Housing-Because lead-based paint is the most important source of lead exposure for young children, the first essential element of primary prevention is implementation of strategies to control lead paint-contaminated house dust and soil and poorly maintained lead paint in housing.[23-25] After 10 or more years of widespread blood lead testing and data collection by CDC-supported state and local agencies, the specific addresses of housing units at which children repeatedly have been identified with elevated BLLs are known to local officials. Two examples are:

- In Detroit, 657 addresses accounted for nearly 1,500 children with BLLs $\geq$20 μg/dL during the last 10 years because the sources of lead were never controlled completely when the initial case occurred. These housing units also probably were the source of lead exposure for several thousand Detroit children with BLLs $\geq$10 μg/dL.[26]

- In Louisville, Kentucky, 35% of children identified with elevated BLLs during the last 5 years resided in 79 housing units; these units represent <0.3% of all housing units in the community.[27]

These experiences are repeated in high-risk communities across the country. The infrastructure needed to identify high-risk housing and to prevent and control lead hazards in such housing is largely in place. Established firms certified in lead hazard evaluation and control now exist in most communities, as do other skilled trades people trained in lead-safe work practices necessary during routine maintenance and painting. Systematic identification and reduction of residential lead sources, particularly in old, poorly maintained housing where children with elevated BLLs are known to have lived, combined with periodic monitoring of housing conditions to detect new deterioration and resultant lead hazards will prevent lead exposure to children in the future and break the cycle of repeated cases of elevated BLLs.

Other steps critical to success in controlling lead hazards in housing and preventing lead exposure in the future are 1) enforcement of lead safety and housing code requirements to ensure good property maintenance; 2) widespread adoption of lead-safe work practices to control, contain, and clean up lead dust during painting and remodeling projects; and 3) periodic monitoring of housing conditions to detect new deterioration and resultant lead hazard.

Nonessential Uses of Lead-Because areas of the United States report that as many as 35% of children identified with elevated BLLs have been exposed to items decorated or made with lead, in some cases resulting in life-threatening BLLs,[28] the second crucial element of a primary prevention strategy is identification and restriction or elimination of nonessential uses of lead, particularly in both imported and domestically manufactured toys, eating and drinking utensils, cosmetics, and traditional medicines. This effort requires identifying communities where cultural practices and traditional medicines may put children at risk and incorporating lead poisoning prevention activities into health and community services that reach families at high risk for lead exposure from nonpaint sources.[7] The 2010 health objective cannot be achieved without a more systematic approach that, at a minimum, allows identification of lead-contaminated items and prohibits their sale before children are exposed. Ultimately, all nonessential uses of lead should be eliminated.

RECOMMENDATIONS

Changes in the Focus of CDC-Funded Programs

To achieve these goals, CDC is focusing on eliminating childhood lead poisoning by preferentially funding programs to provide lead-related services for communities and populations with large numbers of children at high risk for lead exposure. The cooperative agreements with 42 state and local health departments funded for 2003-2006 emphasize the importance of primary prevention and require funded state and local programs to work aggressively to develop and implement the necessary partnerships, programs and activities. CDC requires its state and local partners to undertake a strategic planning process, which includes gathering input from housing professionals, pediatric health-care providers, advocacy groups, parents of children with elevated BLLs, and others interested in preventing lead poisoning in children. These strategic plans, developed by local partners to respond to local conditions, drive primary prevention activities for the 3-year grant cycle.

Progress toward eliminating childhood lead poisoning can be measured only by ongoing surveillance of BLLs in childhood populations where the risk for exposure is high, as well as continued monitoring of population-based BLLs through the National Health and Nutrition Examination Survey.[29] CDC's role in supporting state and local efforts and providing technical assistance to improve data management and reporting is essential to these activities.

Recommendations to Federal, State, and Local Government Agencies

Achieving the Healthy People 2010 objective to eliminate childhood lead poisoning requires collaboration by many different federal, state and local agencies. Many of the roles and responsibilities for federal partners in the elimination effort are detailed in the report of the President's Task Force on Environmental Health Risks and Safety Risks in Children.[29] However, all levels of government share responsibility for primary prevention of childhood lead poisoning. Government agencies have the ability, through legislative and enforcement actions to spearhead prevention efforts and articulate clear public health goals and strategic priorities at the federal, state, and local government levels.

Federal agencies should:

1. Support and disseminate information about, and adequately fund, programs and interventions that will lead to full implementation of primary prevention.

2. Expand financial resources for permanent measures to control or eliminate residential lead hazards.

3. Monitor and enforce regulations controlling lead content of various environmental media, including air, water, and soil.

4. Identify populations in which the risk for exposure to nonpaint sources of lead is high, and develop strategies to minimize the risk.

5. Develop and implement regulatory and voluntary strategies to control nonessential uses of lead, particularly in items that are easily accessible to young children, such as toys, jewelry, eating and drinking utensils, traditional remedies, and cosmetics.

6. Evaluate the effectiveness of primary prevention activities in reducing lead exposure and eliminating childhood lead poisoning, particularly in areas where the risk for lead poisoning is substantially higher than for the general U. S. childhood population.

7. Develop new mathematical models of lead exposure or modify existing models, e.g., the Integrated Exposure Uptake and BioKinetic (IEUBK) Model for lead in children, currently used to establish thresholds for lead exposure in consumer products and areas with pervasive lead contamination. The exposure modeling should predict the magnitude of the increase in BLLs in a child as a result of exposure to a specific lead source rather than the probability of a BLLs ≥ 10 µg/dL.

State and local agencies should:
1. Update or establish and enforce regulatory requirements for lead safe housing that link lead safety to the housing and/or sanitary code.

2 Require that properties that have undergone lead paint abatement or substantial renovation to lead painted surfaces meet the EPA dust clearance testing prior to re-occupancy. Require dust testing in all cases where public health agencies have ordered paint repair, particularly in the homes of children already identified with elevated BLLs.

3 Promote broad use of lead-safe work practices for routine painting and maintenance projects in older homes, and make training in such practices widely available at low or no cost to painters, remodelers, landlords, and maintenance workers.

4. Establish formal agreements among health, social services, housing, and legal agencies to increase the sharing of data, educational information, violations, and success stories.

5. Provide information to caregivers about temporary measures that can reduce lead exposure as described in *Managing Elevated Blood Lead Levels Among Young Children,*[6] as well as information and referral for permanent abatement services.

Recommendations to Health-Care Providers and Community-Based Health and Social Service Agencies

CDC recommends that health care providers continue their traditional role of providing anticipatory guidance as part of routine well-child care, assessing risk for exposure to lead, conducting blood lead screening in children, and treating children identified with elevated BLLs. In addition health-care and social service providers are urged to expand their roles. They should keep abreast of research data that clarify the relationship between lead exposure and neurocognitive development in children. They also can strongly advocate for children and foster lead exposure prevention by helping facilitate implementation of the specific strategic plans to eliminate childhood lead poisoning in their local and state communities. Health-care and social service providers are highly effective child advocates, and their active participation in the process provides the expertise and leadership needed to reach this goal. Health-care and social service providers should:

1. Provide culturally appropriate education to all pregnant women and to families with young children about the principal sources of lead and ways to reduce exposure.

2. Target outreach, education, and screening programs to populations with the greatest risk for lead exposure.

3. Become aware of, and actively support, lead poisoning elimination efforts in the community.

4. Express concern to federal, state, and local policy and decision makers that children live in a lead safe environment and actively support legislation and regulatory initiatives. Advocate for lead-safe, affordable housing by supporting appropriate legislation.

5. Become aware of and comply with lead screening policies issued by Medicaid or state and local health departments.

6. Ensure training of staff members engaged in housing renovation or rehabilitation in lead-safe work practices.

CONCLUSION

The Healthy People 2010 objective to eliminate BLLs >10 µg/dL in children is within our grasp. Research to further characterize and isolate the harmful effects of lead associated with various BLLs will help answer remaining questions and further refine the public health response. However, the approach needed is clear: identify and address existing lead hazards before children are exposed, otherwise hundreds of thousands of children will be placed at risk needlessly. The overall reduction of lead in the environment will benefit all children.

REFERENCES

[1] US Department of Health and Human Services. Healthy People 2010: Understanding and improving health. Washington, DC: US Department of Health and Human Services; 2000. Available at http://www.healthypeople.gov. Accessed 2005 Mar 5.

[2] Meyer PA, Pivetz T, Dignam TA, Homa DM, Schoonover J, Brody D. Surveillance for elevated blood lead levels among children—United States, 1997–2001. MMWR Surveill Summ 2003;52:1–21.

[3] Brown MJ, Shenassa E, Tips N. Small area analysis of risk for childhood lead poisoning. Washington, DC: Alliance to End Childhood Lead Poisoning; 2001.

[4] CDC. Blood lead levels in young children—United States and selected states, 1996–1999. MMWR 2000;49;1133–7.

[5] Kaufmann RB, Clouse TL, Olson DR, Matte TD. Elevated BLLs and blood lead screening among US children aged one to five years: 1988–1994. Pediatrics 2000;106:1–7.

[6] CDC. Managing elevated blood lead levels among young children: Recommendations from the Advisory Committee on Childhood Lead Poisoning Prevention. Atlanta: US Department of Health and Human Services; 2002. Available at http://www.cdc.gov/nceh/lead/CaseManagement/caseManage_main.htm. Accessed 2005 Mar 7.

[7] Rogan WJ, Dietrich KM, Ware JH, Dockery DW, Salganik M, Radcliffe J, et al. The effect of chelation therapy with succimer on neuropsychological development in children exposed to lead. N Engl J Med 2001;344:1421–26.

[8] Rhoads GG, Ettinger AS, Weisel CP, Buckley TJ, Goldman KD, Adgate J, et al. The effect of dust lead control on blood lead in toddlers: A randomized trial. Pediatrics 1999;103:551–5.

[9] Tohn ER, Dixon SL, Wilson JW, Galke WA, Clark CS. An evaluation of one-time professional cleaning in homes with lead-based paint hazards. Appl Occup Environ Hyg 2003;18:138–43. Ê

[10] Weitzman M, Aschengrau A, Bellinger D, Jones R, Hamlin JS, Beiser A. Lead-contaminated soil abatement and urban children's blood lead levels. JAMA. 1993;269:1647–54.

[11] Dietrich KN, Ware JH, Salganik M, Radcliffe J, Rogan WJ, Rhoads GG, et al. Effect of chelation therapy on the neuropsychological and behavioral development of lead-exposed children after school entry. Pediatrics 2004;114:19–26.

[12]Brown MJ, McLaine P, Dixon S, Simon P. A randomized community-based trial of home visiting to reduce blood lead levels in children. Pediatrics. In press 2005.

[13]Moel DI, Sachs HK, Drayton MA. Slow and natural reduction in blood lead level after chelation therapy for lead poisoning in childhood. Am J Dis Child 1986;140:905–8.

[14]Roberts JR, Reigart JR, Ebeling M, Hulsey TC. Time required for blood lead levels to decline in nonchelated children. Clin Toxicol 2001;39:153–60.

[15]Manton WI, Angle CR, Stanek KL, Reese YR, Kuehnemann TJ. Acquisition and retention of lead by young children. Environ Res Sect A 2000;82:60–80.

[16]Clark S, Grote JA, Wilson J, Succop P, Chen M, Galke W, et al. Occurrence and determinants of increases in blood lead levels in children shortly after lead hazard control activities. Environ Res 2004;96:196–205.

[17]CDC. Preventing lead poisoning in young children. Atlanta: US Department of Health and Human Services; 1991.

[18]Bernard SM, McGeehin MA. Prevalence of blood lead levels ≥ 5 µg/dL among US children 1 to 5 years of age and socioeconomic and demographic factors associated with blood lead levels 5 to 10 µg/dL, Third National Health and Nutrition Examination Survey, 1988–1994. Pediatrics 2003;112:1308–13.

[19]CDC. Strategic plan for the elimination of childhood lead poisoning. Atlanta: US Department of Health and Human Services; 1991.

[20]Grosse S, Matte T, Schwartz J, Jackson R. Economic gains resulting from the reduction in children's exposure to lead in the United States. Environ Health Perspect 2002;110:563–9.

[21]Housing and Community Development Act. 1992. Title X Residential Lead-based Paint Hazard Eduction Act. Pub. L. 102-550, 42 U.S. 4822.

[22]CDC. Preventing lead exposure in young children: A housing-based approach to primary prevention of lead poisoning. Atlanta: US Department of Health and Human Services; 2004.

[23]Bornschein RL, Succop PA, Krafft KM, Clark CS, Peace B, Hammond PB. Exterior surface dust lead, interior house dust lead and childhood lead exposure in an urban environment. In: Trace Substances in Environmental Health; Proceedings of the University of Missouri's annual Conference on Trace Substances in Environmental Health. St. Louis: University of Missouri; 1987;20:322–32.

[24]Lanphear BP, Matte TD, Rogers J, Clickner RP, Dietz B, Bornschein RL, et al. The contribution of lead-contaminated house dust and residential soil to children's blood lead levels. A pooled analysis of 12 epidemiologic studies. Environ Res 1998;79:51–68.

[25]Manton WI, Angle CR, Stanek KL, Reese YR, Kuehnemann TJ. Acquisition and retention of lead by young children. Environ Res 2000;82:60–80.

[26]Wendland-Bowyer W. Worst Michigan Neighborhoods: Lead-poisoned blocks pinpointed. Detroit Free Press 2003 July 29;1.

[27]Meyer PA, Staley F, Staley P, Curtis J, Blanton C, Brown MJ. Improving strategies to prevent childhood lead poisoning using local data. Int J Hyg Environ Health In Press 2005.

[28]CDC. Brief report: Lead poisoning from ingestion of a toy necklace—Oregon, 2003. MMWR 2004;53:509–11.

[29]President's Task Force on Environmental Health Risks and Safety Risks to Children. Eliminating childhood lead poisoning: A federal strategy targeting lead paint hazards. Washington, DC: President's Task Force on Environmental Health Risks and Safety Risks to Children; 2004.

APPENDIX

A REVIEW OF EVIDENCE OF ADVERSE HEALTH EFFECTS ASSOCIATED WITH BLOOD LEAD LEVELS <10 µg/dL IN CHILDREN

Reported by

Work Group of the Advisory Committee on
Childhood Lead Poisoning Prevention

to

CENTERS FOR DISEASE CONTROL AND PREVENTION
National Center for Environmental Health

AUGUST 2005

Appendix Table of Contents

List of Figures

List of Tables

EXECUTIVE SUMMARY Ê

In March 2002, the Centers for Disease Control and Prevention (CDC) Advisory Committee on Childhood Lead Poisoning Prevention (ACCLPP) established a work group (WG) to review the available evidence of possible health effects of blood lead levels (BLLs) of below 10 micrograms per deciliter (µg/dL), the level of concern currently established by CDC. The WG was charged with designing and following a rigorous protocol to review studies of the health effects of lead exposure at very low BLLs. The workgroup intended to focus on studies of the effects of peak BLLs at <10 µg/dL in children never known to have a BLL exceeding 10 µg/dL. However, there are relatively few such studies and the workgroup decided to review the larger number of studies that could indirectly support or refute the existence of a threshold near 10 µg/dL. Although the workgroup members were the primary authors of this report, the ACCLPP reviewed the document and it was revised based on their comments. The majority of ACCLPP members accepted the findings of the report, with two members dissenting.

Methods

The following criteria were used for selecting relevant studies to review:

- BLLs were measured using graphite furnace atomic absorption spectrometry (GFAAS) or anodic stripping voltammetry (ASV);

- the study was published in English;

- for studies in which IQ or General Cognitive Index (GCI) was a measured outcome, an assessment of the association between BLLs in children and IQ or GCI was included; and

- for studies in which IQ or GCI was not a measured outcome, an assessment of the association between BLLs in children and a specified health outcome was included.

For each relevant study, a structured abstraction was performed that captured the following:

- study location and sample size;

- age at which BLL and cognitive or health outcome was measured;

- the distribution of BLLs (mean or other measure of central tendency and variance) and percentage of participants with BLL <10 µg/dL;

- crude and adjusted regression coefficients relating BLLs to outcome;

- other measures of association (e.g., correlation coefficients); and

- model type and covariates included in adjusted models.

When reviewing the evidence, including indirect evidence from IQ studies, the workgroup considered both alternate explanations for study findings and potential effect of residual confounding.

Conclusions

The main conclusions reached by the WG are summarized as follows.

1. **Does available evidence support a negative association between measured BLLs <10 µg/dL and children's health?**

    - The overall weight of available evidence supports an inverse (negative) association between BLLs <10 µg/dL and the cognitive function of children.

    - A steeper slope in the dose-response curve was observed at lower rather than higher BLLs.

    - The available evidence has important limitations, including the small number of directly relevant cohort studies and the inherent limitations of cross-sectional studies (i.e., the lack of data regarding both BLLs earlier in life and key covariates).

    - For health endpoints other than cognitive function (i.e., other neurologic functions, stature, sexual maturation, and dental caries), consistent associations exist between BLLs <10 µg/dL and poorer health indicators.

2. **Are the observed associations likely to represent causal effects of lead on health?**

    - Though not definitive, the available evidence supports the conclusion that the observed associations between BLLs <10 µg/dL and cognitive function are caused, at least in part, by lead toxicity.

    - The strength and shape of the causal relationship are uncertain because of limitations of the available evidence.

- The health effects of lead are uncertain in individual children who have BLLs measured at a single point in time. Thus, scientific evidence does not provide a basis for classifying individual children with BLLs <10 µg/dL as "lead poisoned," as the term is used in the clinical setting.

- The greatest source of uncertainty in evidence concerning the relationship between BLLs <10 µg/dL and children's cognitive function is the potential for residual confounding, especially by socioeconomic factors.

- The available data for health endpoints other than cognitive function, taken mostly from cross-sectional studies, are limited; therefore, firm conclusions concerning causation can not be made.

Future Research Needs

The WG identified the following research needs to address gaps in the existing base of evidence and to allow for more definite conclusions about the strength and shape of the causal relationship.

- Prospective observational studies designed to minimize the chance of residual confounding.

- Randomized trials to test interventions designed to reduce BLLs <10 µg/dL and assess the impact on children's cognitive development.

- Animal and in vitro studies to identify mechanisms of lead toxicity at low BLLs that could explain the observed steeper slope at lower compared with higher BLLs.

A Review of Evidence of Adverse Health Effects Associated with Blood Lead Levels <10 µg/dL in Children

Reported by

A Work Group of the CDC Advisory Committee on Childhood Lead Poisoning Prevention on Health Effects of Blood Lead Levels <10 µg/dL in Children

Work Group Members

Michael Weitzman, MD
Work Group Chair
Center for Child Health Research
University of Rochester

Joel Schwartz, PhD
Department of Environmental Health
Harvard School of Public Health

David Bellinger, PhD
Neuroepidemiology Unit
Children's Hospital
Harvard Medical School

David A. Savitz, PhD
Department of Epidemiology
University of North Carolina School
of Public Health

Carla Campbell, MD
Division of General Pediatrics
The Children's Hospital of Philadelphia

Patrick J. Parsons, PhD
Wadsworth Center for
Laboratories and Research
New York State Department
of Health

Betsy Lozoff, MD
Center for Human Growth
and Development
University of Michigan

Kimberly M. Thompson, ScD
Department of Health Policy
and Management
Harvard School of Public Health

Birt Harvey, MD
Pediatrician
Palo Alto, California

This review was prepared for the work group by the following:

Thomas D. Matte, MD, MPH
David Homa, PhD
National Center for Environmental Health
Centers for Disease Control and Prevention
and
Jessica Sanford, PhD
Alan Pate
Battelle Memorial Institute

Abbreviations and Acronyms

AAS	atomic absorption spectrometry
ACCLPP	Advisory Committee on Childhood Lead Poisoning Prevention
ALAD	amino levulinic acid dehydratase
ALAU	urinary amino levulinic acid
ASV	anodic stripping voltammetry
ATSDR	Agency for Toxic Substances and Disease Registry
BLL	blood lead level
EBLL	elevated blood lead level
EP	erythrocyte protoporphyrin
EPA	Environmental Protection Agency
ETAAS	electrothermal atomization techniques based on the graphite furnace
ETS	environmental tobacco smoke
FEP	free erythrocyte protoporphyrin
GCI	General Cognitive Index

GFAAS	graphite furnace atomic absorption spectrophotometry
HOME	Home Observation for Measurement Environment
ICP-MS	inductively coupled plasma mass spectrometry
ID-MS	isotope dilution mass spectrometry
MCV	mean corpuscular volume
MDI	Mental Developmental Index of the Bayley Scales of Infant Development for children
MeHg	methylmercury
MSCA	McCarthy Scales of Children's Ability
NCCLS	National Committee for Clinical Laboratory Standards
NCEH	National Center for Environmental Health
NHANES	National Health and Nutrition Examination Survey
NMDA	N-methyl-D-aspartate
PbB	Blood lead
PCAACN	Practice Committee of the American Academy of Clinical Neuropsychology
PKC	protein kinase C, a calcium dependent enzyme

QA/QC	quality assurance/quality control
SES	socioeconomic status
U-RBP	urinary retinal binding protein
WG	work group
WISC-R	Wechsler Intelligence Scale for Children—Revised
WISC-III	Wechsler Intelligence Scale for Children—Third Edition
WRAT	Wide Ranging Achievement test arithmetic and reading scores
ZPP	zinc protoporphyrin

Background

Charge to the Work Group

In March 2002, the Advisory Committee on Childhood Lead Poisoning Prevention agreed to establish a work group (WG) to review evidence of possible health effects of lead at blood lead levels less than 10 micrograms per deciliter (µg/dL), currently the threshold for defining an elevated blood lead level according to CDC guidelines (CDC 1991). The work group was charged as follows:

> "In October 1991, the Centers for Disease Control and Prevention issued Preventing Lead Poisoning in Young Children. This document heralded a change in the definition of the level for intervention for children with elevated blood lead levels (EBLLs) from a lead level of 25 µg/dL to 10 µg/dL. The report explained that this change was due to new data that indicated significant adverse effects of lead exposure in children at levels once thought to be unassociated with adverse effects. The 1991 document identified a goal to reduce children's blood lead levels below 10 µg/dL. Interventions for individual children were recommended at levels of 15 µg/dL and above.
>
> Research findings published and disseminated since October 1991 suggest that adverse effects from lead exposure and toxicity occur at blood lead levels below 10 µg/dL. Some studies suggest that some effects may be greater at blood lead levels (BLLs) below 10 µg/dL than at higher BLLs. Such research findings raise concerns about the inability to control lead exposure with conventional methods and lend credence to the importance of primary prevention measures to prevent lead exposure to children.
>
> The work group will be convened by the Advisory Committee on Childhood Lead Poisoning Prevention to review the existing evidence for adverse effects of lead exposure and toxicity on children at very low blood lead levels and to focus on effects at levels of 10 µg/dL and below. Rigorous criteria will be established for the literature review. The work group will then create, in conjunction with the committee, a summary of the evidence for publication."

Scientific and Public Health Context for the WG Review

Prior reviews that compiled the extensive evidence from *in vitro*, animal, and human studies established lead as a multi-organ toxicant, including studies showing health effects at BLLs near 10 µg/dL (ATSDR 1999; WHO 1995; USEPA 1986). The published studies include a large body of literature establishing that

1

lead is a developmental toxicant and that harmful effects of lead on children's development can occur without clinical signs, symptoms, or abnormal routine laboratory tests. In addition, a growing number of studies suggest that BLLs prevalent in the general population are associated with adverse health effects in adults and in the offspring of pregnant women. Finally, in more recent years, bone-lead levels measured by x-ray fluorescence have been used in epidemiologic studies as a measure of cumulative lead exposure. Although these were not considered in this review, a number of studies showing inverse relations of bone-lead level to health in general population samples (e.g., Cheng et al. 2001) add further evidence that cumulative lead exposure may be harmful to health at typical background exposure levels for the population in the United States.

The observation that available epidemiologic evidence does not demonstrate a threshold below which no effect of lead is possible is not new. A review prepared for a 1986 workshop on lead exposure and child development stated, "There is little evidence for a threshold or no-effect level below which the lead/IQ association is not found. IQ deficits have been reported in studies where the mean lead level is 13 µg/dL (Yule et al. 1981) and similar deficits in the Danish study where the primary measure was tooth lead, but BLLs are lower at around 7 µg/dL." (Smith 1989) A review, meta-regression, and reanalysis of existing data (Schwartz 1994) reached essentially the same conclusion. Available data did not suggest a threshold below which no association between BLLs and intelligence in young children was evident. Recent studies (Canfield et al. 2003; Lanphear et al. 2000; Moss et al. 1999; Wu et al. 2003) provided more direct evidence of an association between BLLs and adverse health effects in the domains of cognitive function, neurologic function, growth, dental caries, and onset of puberty at levels well below 10 µg/dL. Thus, a reexamination of this issue is in order.

As evidence from experimental animal studies and human epidemiologic studies has grown, CDC has lowered the BLL considered elevated for the purpose of interpreting clinical test results of an individual child (Table 1). CDC guidelines also have provided criteria for identifying children who have more severe manifestations of lead toxicity and/or a higher risk of lead-related sequellae. For example, CDC's 1975 and 1978 guidelines defined clinical "lead poisoning" on the basis of BLLs, symptoms, and/or levels of erythrocyte protoporphyrin (EP) or other indicators of lead-related biochemical derangements. CDC's 1985 guidelines used the terms "lead toxicity" and "lead poisoning" interchangeably to refer to BLLs ≥25 µg/dL with EP ≥35 µg/dL. However, the guidelines acknowledged that "lead poisoning" is generally understood for clinical purposes to refer to episodic, acute, symptomatic illness from lead toxicity. CDC's 1985 guidance also cautioned that blood lead thresholds established to guide follow-up and treatment for individual children "should not be interpreted as implying that a safe level of blood lead has been established." In 1991, CDC guidelines more directly acknowledged the difficulty in assigning terms to specific ranges of BLLs given the different settings in which BLLs are interpreted and given that manifestations of lead toxicity occur along a continuum: "It is not possible to select a single number to define lead poisoning for the various purposes of all of these groups [e.g. clinicians, public health officials, and policy makers]" (CDC 1991). These guidelines also noted, "Some [epidemiologic] studies have suggested harmful effects at even lower levels [than a BLL of 10 µg/dL]."

In addition to these changes in criteria used to evaluate blood lead test results for individual children, recent analyses by the U.S. Department of Housing and Urban Development (HUD 1999) and the U.S. Environmental Protection Agency (USEPA 2000) to support the development of regulations governing lead exposure have assumed that the relation of increasing blood lead to decrements in children's IQ extends to BLLs <10 µg/dL.

As BLLs considered elevated have fallen, measures to reduce or remove lead from a number of sources, including gasoline, soldered food and beverage containers, paint, drinking water, and industrial emissions have resulted in a dramatic decline in BLLs in the United States since the mid-1970s (Pirkle et al. 1994). The Second National Health and Nutrition Examination Survey (NHANES II) conducted from 1976 to 1980 demonstrated that, among U.S. children ages 6 months through 2 years, 84% of white children and more than 99% of black children had BLLs ≥10 µg/dL, and the median BLLs were 14 and 19 µg/dL, respectively (Mahaffey et al. 1982). A decline in BLLs during the course of that survey was noted, paralleling the falling consumption of leaded gasoline (Annest et al. 1983). A continued decline in BLLs was evident in subsequent NHANES surveys (Pirkle et al. 1994; NCEH 2003) and in clinical blood-lead test data compiled by state and local health agencies (Hayes et al. 1994; CDC 2003). Nationally, it is estimated that by 1999-2000, the prevalence of BLLs ≥10 µg/dL among children 1 to 5 years of age had fallen to 2.2% and the median level to 2.2 µg/dL (NCEH 2003).

Although these reductions in lead exposure represent great progress, scientific advances have shed light on harmful effects of lead at levels of exposure once thought safe. In addition, industrial activity has widely dispersed lead in the environment from naturally occurring deposits. As a result, even at the lower exposure levels that prevail today, typical body burdens of lead are likely to be much higher than those present in pre-industrial humans, which by one estimate corresponded to a BLL of 0.016 µg/dL (Smith et al. 1992). Therefore, the potential for additional subclinical adverse effects of lead from currently prevailing exposures deserves careful study. Finally, although falling BLLs have benefited all demographic groups (Pirkle et al. 1994), stark demographic and geographic disparities continue to reflect the historic pattern; the risk of elevated BLLs in communities where poverty and older (i.e., that built before 1950) housing are prevalent remains several fold higher than the national average (Lanphear et al. 1998).

Review Methods

Scope and Approach

Given the charge to the work group and the scientific and public health context, the WG did not attempt a comprehensive review of all evidence relating lead exposure to health. Instead, the WG set out to answer the following questions:

1. Does available evidence support negative associations between health indicators and children's blood lead levels measured <10 µg/dL? Ê

2. Are the observed associations likely to represent a causal effect of lead on health?

To address these questions, the WG established criteria (see Methods) for published studies that would address the first question. In addition, the work group identified issues relevant to making causal inference from any observed associations. Identifying such issues is an essential step in interpreting evidence relevant to the WG charge. Human studies to assess potential health effects of environmental toxicants, such as lead, are usually observational in design (i.e., the health status of participants is related to some measure of exposure, dose, or body burden that varies on the basis of environmental factors and not experimental manipulations by the investigator). For ethical reasons, the limited number of human experimental studies that have evaluated causal relations between toxicant exposures and health usually have involved attempts to reduce exposure or body burden and assess the impact on health status. Such studies of lead-exposed children are rare, and to date none have focused on children with BLLs <10 µg/dL.

Observational studies have inherent limitations—not specific to studies of lead toxicity—with the potential to produce biased results. Biases from observational studies can obscure true causal effects of toxicant exposures or produce associations between toxicant exposures and health status when no causal relation is present. Thus, statistical associations from individual observational studies or multiple studies subject to similar biases cannot establish causal relationships; additional, non-statistical criteria may be used to evaluate such evidence. Although causal criteria have been stated in various ways, the Surgeon General's Report on Smoking and Health (U.S. Public Health Service 1964) provides a useful set of criteria. They include:

- The consistency of the association. Is a similar association observed across studies with varying methods and populations?

- The strength of the association. The strength of an association is the extent to which the risk of a disease or a measure of health status varies in

4

relation to exposure and can be expressed, for example, as a relative risk or regression coefficient. It is distinguished from the statistical significance of an association that reflects both the strength of the association and the sample size. An additional criterion, specificity of the association, is closely related to strength of the association and is considered less important in the context of multifactorial health conditions.

- The temporal relationship of associated variables. Does the hypothesized causal exposure occur before the health outcome associated with it?

- Coherence of the association. Is the observed association consistent with other relevant facts including, for example, experimental animal studies and the descriptive epidemiology of the health condition under study?

The application of these criteria does not provide a clear demarcation for concluding definitive proof of causation versus inadequate evidence. Rather, the more the available evidence meets these criteria, the greater the confidence in causal inference about an association. Consistent with these criteria, the WG identified several issues specifically relevant to inferring causality from associations (or the lack of associations) of BLLs to health measures observed in studies of low-level lead exposure. These potential biases are not unique to studies of children with BLLs ≤10 µg/dL (e.g., Smith 1989). However, the larger number of human and experimental animal studies (including primate studies) and the nature of observed health effects associated with higher BLLs have conclusively demonstrated the adverse health effects of lead. However, far fewer studies of possible health effects of BLLs <10 µg/dL have been conducted, and the relative importance of some sources of bias may be greater at these lower levels. Therefore, the work group considered several issues in interpreting the findings of available studies (see Discussion).

At the time of the WG's review, a consortium of investigators from several longitudinal studies of lead exposure and cognitive function in children were conducting a reanalysis of data pooled from these studies. These studies included serial measures of blood lead level, cognitive function, and a large number of potential confounders, thus providing stronger evidence than is available from cross-sectional studies. A focus of the pooled reanalysis involved studying the shape of the association between postnatal lead exposure at low levels and measured IQ (B. Lanphear, personal communication, 2003). The WG reviewed published reports from individual cohort studies from which data were pooled, but the final results of this pooled re-analysis were not available for inclusion in the WG report. [Note: Results of the pooled renanlysis were published (Lanphear et al 2005) after the WG finished its review.]

Because of the nature of its charge, the WG did not develop policy recommendations or address questions relevant to such recommendations. Such policy decisions and questions will, if appropriate, be considered by the full ACCLPP after reviewing the findings of this report.

Criteria for Relevant Studies

The WG initially considered, then rejected, limiting its review to studies for which published results provide direct comparisons between children with varying BLLs <10 µg/dL. Such a review would include a relatively small number of studies. Instead, the group decided that the larger number of studies that have included IQ as an outcome could, collectively, indirectly support or refute the existence a threshold near 10 µg/dL for the blood lead–IQ association. The rationale for this approach is based on that used in a review and metaregression reported by Schwartz (1994) and outlined in the following paragraphs.

Suppose that, hypothetically, a threshold exists near 10 µg/dL, above which mean IQ decreases linearly with increasing blood lead, with a slope equal to x and below which mean IQ is not associated with blood lead[1] (see Figure 1—Hypothesized "true" relation A). In studies of children who have BLLs <10 µg/dL, estimated slopes would be, in the absence of sampling error, equal to 0. For studies in which all children have BLLs above the threshold, the estimated slope would be, again ignoring sampling error, equal to x. For studies in which some children have blood lead above and others below the threshold, estimated slopes will vary between 0 and x as shown in Figure 1. Thus, if regression coefficients estimating the IQ–blood lead slope are less negative (approaching 0) in populations with lower mean BLLs than in populations with higher mean levels, a threshold near 10 would be suggested. The absence of such a trend or an increase in slopes with decreasing mean blood lead level of the population studied would provide evidence against such a threshold and instead support "true" relations B and C, respectively. This ideal hypothetical case presumes that effect sizes from the studies compared are based on models that correctly specify the form of the BLL–IQ relation and that factors that might modify the relation do not vary across studies.

Because of this approach, studies that assessed the association between blood lead and measured IQ were included in this review, even if the published results did not examine blood lead–IQ associations limited to BLLs <10 µg/dL (as was true in most cases). An additional reason for considering studies that measured IQ was the relationship of IQ to other outcomes of policy and public health importance including educational success and earnings potential (Grosse et al. 2002). Because the McCarthy Scales of Children's Ability (MSCA) General Cognitive Index (GCI) was used in a number of studies to measure cognitive function in preschool children and because GCI and IQ scores have similar distributions, studies using GCI as an outcome were also included in this review.

---

[1]The concept of a threshold existing for the population makes little sense toxicologically since even if individual thresholds exist, these are likely to vary. Nonetheless, the threshold concept plays a major role in regulatory toxicology, and it only becomes clear in cases like lead that such constructs can be highly problematic.

The following criteria were used to select relevant studies to review for this report:

1. Blood lead levels were measured using graphite furnace atomic absorption spectrophotometry (GFAAS) or anodic stripping voltametry.

2. The study was published in English.

In addition,

For studies in which IQ or GCI was a measured outcome, the

study analyses included an assessment of the association between BLLs measured in children and IQ or GCI.

For studies in which IQ or GCI was not a measured outcome, the

study analyses included an assessment of the association between BLLs <10 µg/dL measured in children and a health outcome. The assessment could either be formal (e.g., non-linear modeling, linear modeling restricted to populations with all or at least 95% of children having BLLs <10 µg/dL, statistical comparison of two or more sub-groups with BLLs <10 µg/dL) or informal (e.g., graphical display of results permitting visual assessment of blood lead–outcome relation in the range <10 µg/dL).

Literature Search

To identify potentially relevant articles, a comprehensive report published by the Agency for Toxic Substances and Disease Registry (ATSDR 1999) was reviewed first to identify cited articles that related to low-level lead exposure in children. The list of potentially relevant citations identified in the ATSDR report was supplemented by three computerized literature searches, using Dialog® to search Medline, Toxfile, and other bibliographic databases. Search terms (see Appendix A) were chosen to identify articles reporting on blood lead measurements and one or more domains of health related to lead exposure including neurodevelopment, cognitive function, intelligence, behavior, growth or stature, hearing, renal function, blood pressure, heme synthesis, hematopoiesis, and Vitamin D metabolism. The first search spanned articles published from 1995 through 2002 and indexed as of September 2002, the month and year that the initial search was performed. The second search was performed in April 2003 and spanned the period 2002 through the search date. A third search, spanning the years 1990 through 1996, was performed when a relevant article not cited in the ATSDR toxicological profile was identified by one of the work group members. In addition, potentially relevant articles were identified by work group members and through citations in articles identified previously.

Abstracts were reviewed initially. If they were ambiguous or if they suggested the article was relevant, full articles were checked for relevance. Articles deemed relevant were abstracted for this report. Appendix A summarizes the number of possibly relevant references identified, full articles checked for relevance, and relevant articles abstracted and considered in the review.

Structured Abstracts

For each relevant study, a structured study abstraction was performed that captured the following information: the study location, sample size, age at which blood lead was measured, age at which the outcome was measured, available information about the blood lead distribution (including mean or other measure of central tendency, variance, and percent of participants with BLLs <10 µg/dL), the crude and adjusted regression coefficients relating blood lead to outcome (if available), the type of model fit (linear, log linear, or other), and the covariates included in the adjusted model. If regression coefficients were not available, other measures of association reported (e.g., correlation coefficients) were noted. Because some studies fit multiple blood lead-outcome models (e.g., cohort studies with blood lead and IQ measured at multiple ages), relevant information about each model estimated was abstracted. For IQ studies, covariates measured and not included in adjusted models were recorded when available.

Review of Cohort Study Methods

Among relevant published results were those from cohort studies specifically designed and conducted to study the relation of BLLs to children's cognitive function and other health outcomes. Because these studies had the strongest and best-documented study designs for this review, methods used for blood lead measurement and neuropsychological assessment were summarized for these studies. This information was collected from published studies; in some cases, the studies were supplemented by information provided through correspondence with the investigators.

Results

Studies Relating Postnatal Lead Exposure to IQ or General Cognitive Index

Studies in which IQ or GCI were measured as an outcome and other criteria were met included 23 published reports from 16 separate study populations. Results from these studies are summarized in Table 2 (full scale IQ and GCI), Table 3 (performance IQ), and Table 4 (verbal IQ). Within each table, results are grouped according to the age at blood-lead measurement and age at outcome measurement;

each grouping displays results sorted according to the measure of central tendency of the blood-lead distribution. Because some studies used linear models (BLLs were untransformed) and some used log-linear models (BLLs were log transformed), estimated regression coefficients were, when possible, used to calculate the estimated change in IQ or GCI corresponding to a blood lead increase from 5 to 15 µg/dL to allow for comparisons across studies. Covariates included in adjusted models are grouped into several broad domains that were addressed in most of the published studies.

Among studies that provided results for the size and direction of the associations between BLLs and Full Scale IQ or GCI, regardless of statistical significance, a majority revealed that both crude and adjusted associations were consistent with an adverse effect—IQ decreases with increasing levels of blood lead. In most cases, covariate adjustment attenuated, but did not eliminate, these estimated associations. Findings for performance and verbal IQ were similar, with some studies showing stronger associations of lead with performance IQ and others with verbal IQ.

Notable exceptions to this pattern, however, were found. Results from the Cleveland cohort (Ernhart et al. 1987, 1988, 1989) indicated a crude inverse association between blood lead and IQ but no association with covariate adjustment. In the Cincinnati and Boston cohort studies, BLLs measured at or below 12 months of age showed no association or a slightly positive association with covariate-adjusted IQ (Dietrich et al. 1993; Bellinger et al. 1992). Though published results from a cohort study in Costa Rica (Wolf et al. 1994) did not provide the size and direction of the estimated blood lead–IQ slope, unpublished results provided to the WG did show covariate-adjusted IQ increasing with BLL (B. Lozoff, personal communication 2003). The estimated BLL–IQ association in the Kosovo cohort was strengthened substantially with covariate adjustment (Wasserman et al. 1997).

No trend toward attenuation of the association between blood lead and IQ (or GCI) across studies with decreasing average BLLs is evident (Figures 2 and 3). In one of these studies, analyses were presented that provide more direct information concerning the association between BLL and cognitive function at BLLs <10 µg/dL. The steepest estimates of blood lead–IQ slope from the Rochester (Canfield et al. 2003) studies were based on analyses restricted to children whose measured BLL never exceeded 10 µg/dL. The estimated slope was substantially larger than those estimated from the entire study population (9.2 versus 5.3 IQ point reduction in covariate adjusted IQ for BLL increase from 5 to 15 µg/dL). Canfield and his colleagues (2004) also reported a non-linear model supporting a steeper blood lead–IQ slope at lower levels. Though published as a letter to the editor, rather than in a peer-reviewed article (and therefore not included in the structured review), similar findings were reported for a reanalysis of the Boston cohort: a steeper BLL–IQ slope in the population of children whose measured BLLs never exceeded 10 µg/dL, compared with the entire study population (Bellinger et al. 2003).

Most of the published studies included at least one measure of socio-economic status. All of the published results from cohort studies were adjusted for Home Observation for Measurement Environment (HOME) score and birth weight; all except the Costa Rica cohort were adjusted for a measure of maternal intelligence. A reanalysis of the data from the Costa Rica cohort with adjustment for maternal IQ was consistent with the original finding of a non-significant positive blood lead–IQ slope (B. Lozoff, personal communication, 2003). Prenatal exposure to maternal smoking was adjusted in the majority of studies, whereas only in the Port Pirie cohort was a measure of postnatal environmental tobacco smoke exposure included. Iron deficiency anemia was included as a covariate in results from the Costa Rica study, which found no association of lead and IQ; the inverse blood lead–IQ associations in the Rochester study (Canfield et al. 2003) and the Karachi study (Rahman et al. 2002) were adjusted for serum transferrin saturation and hemoglobin, respectively. In addition, in the Kosovo study (Wasserman et al. 1997), alternative models were fit with adjustment for hemoglobin, resulting in no appreciable change in the lead coefficient.

Not all studies reported regression coefficients that could be used to estimate the change in IQ associated with a BLL change of 5 to 15 µg/dL, and the overall pattern of results summarized above and in Figures 2 and 3 might have been altered had regression coefficients been available from all studies. For example, a member of the WG provided results of reanalysis of the data from the Costa Rica cohort, which showed no evidence of an inverse relation of BLL to IQ with adjustment for maternal IQ and other covariates used in the published result (Wolff et al. 1994; B. Lozoff, personal communication, 2003).

Studies of Health Endpoints Other than IQ and CGI

More stringent criteria were required for inclusion of studies in this review if they assessed health endpoints other than general intelligence as measured by IQ or GCI. These studies are summarized in Table 5 and are described in the following paragraphs.

Cognitive Function

Lanphear et al. (2000) analyzed data on BLLs on performance on standardized tests of cognitive function of 4,853 children age 6 through 16 years who were evaluated as part of the NHANES III survey, a multiphasic health interview and examination survey of a stratified probability sample of the U.S. population, carried out from 1988 through 1994. In this population, with a geometric mean BLL of 1.9 µg/dL and 98% of children having BLLs <10 µg/dL, significant inverse relations were found between BLLs and scores on the Wide Ranging Achievement Test (WRAT) arithmetic and reading scores and on the WISC-R block design and digit span subscales. The relationships were strengthened (the slopes became more negative) as analyses were progressively restricted to children with lower BLLs. Stone et al. (2003) reanalyzed the data used by Lanphear et al. (2000). While the results they present are largely consistent with the findings of Lanphear et al., they provided a critique of the validity of the NHANES III data for evaluating lead-related impacts on neuropsychological development in children. Their critique did not provide results that could be summarized in the structured abstract format used in this report, so a discussion of the Stone et al. critique is found in Appendix B.

Other Neurobehavioral Measures and Visual Function

Three reports (Altman et al. 1998; Walkowiak et al. 1998; Winneke et al. 1994) describe the relation of blood lead to several neurobehavioral measures and to visual function assessed in 384 school children 5 to 7 years of age in three cities in Eastern Germany. Blood lead levels were generally low, with a geometric mean of 4.25 µg/dL and 95% of children having BLLs <10 µg/dL. Walkowiak et al. (1998) reported a significant negative association between BLLs and WISC vocabulary subscale scores. Continuous performance test false positive and false negative responses increased with increasing BLLs. Other measures inversely related to BLL included performance on a pattern comparison test, finger tapping speed (Winneke et al. 1994), and visual evoked potential interpeak latency (Altman et al. 1998). Mendelsohn and colleagues (1999) found a 6 point deficit in the Mental Developmental Index (MDI) of the Bayley Scales of Infant Development for children aged 12 to 36 months with BLLs 10-24 µg/dL compared with children who had BLLs <10 µg/dL. A scatterplot of covariate-adjusted MDI versus blood lead suggests the association continues at BLLs <10 µg/dL.

Neurotransmitter Metabolite Levels

Among children ages 8 through 12 years with mean BLL of 3.95 µg/dL, a direct relation of blood lead (PbB) to higher urinary homovanillic acid, a neurotransmitter metabolite, was found for the subset of children with BLLs >5 µg/dL (Alvarez Leite et al. 2002).

Growth

Two studies examined the relation of BLLs <10 µg/dL to somatic growth. Ballew et al. (1999), using the NHANES III data, found that BLLs were inversely related to height and to head circumference among children 1 to 7 years of age. A birth cohort of children in Mexico had BLLs and head circumference assessed every 6 months from 6 to 48 months of age, during which time the median BLL varied from 7 to 10 µg/dL (Rothenberg et al. 1999). Most postnatal blood lead measures were inversely correlated with covariate adjusted head circumference, with the strongest relation found between blood lead at age 12 months and head circumference at 36 months. Kafourou and colleagues (1997) reported a significant negative association between BLL and covariate-adjusted head circumference and height in a population of children with a median BLL of 9.8 µg/dL, with a scatterplot suggesting the relation extends to BLLs <10 µg/dL.

Sexual Maturation

Two studies, both based on analyses of the NHANES III data, found an association between BLLs <10 µg/dL and later puberty in girls. Selevan et al. (2003) found that BLLs of 3 µg/dL, compared with 1 µg/dL, were associated with significant delays in breast and pubic hair development in African American and Mexican girls. The trend was similar, but not significant, for non-Hispanic white girls. Age at menarche was also delayed in relation to higher BLLs, but the association was only significant for African-American girls. Wu et al. (2003) reported similar findings for girls in the NHANES III population but did not stratify the analysis by racial/ethnic group. Compared with BLLs 2.0 µg/dL and below, BLLs of 2.1–4.9 µg/dL were associated with significantly lower odds of attaining Tanner 2 stage pubic hair and menarche; whereas no overall association with breast development was noted.

Dental Caries

In the NHANES III population, the odds of having dental caries, comparing children ages 5–17 years in the middle tertile of the BLL distribution (range of BLLs 1.7–4.1 µg/dL) with those in the lowest tertile, was significantly elevated (odds ratio=1.36, 95% confidence interval 1.01–1.83) (Moss et al. 1999). Gemmel et al. (2002) evaluated the association between BLLs and caries in 6–10 year old children from urban communities in eastern Massachusetts (mean BLL=2.9 µg/dL) and a rural community (mean BLL=1.7 µg/dL) in Maine. They found a significant direct relation of BLL to caries in the former, but not the latter population in which a non-significant decrease in caries' frequency was observed with increasing blood lead. In the urban population, the trend of increasing caries with PbB level was evident comparing children with BLLs of 1, 2, and 3 µg/dL.

Blood pressure and renal function

Among 66-month-old children in Kosovo, a graph depicting adjusted mean systolic and diastolic blood pressure versus BLL showed no consistent trend across 4 groups of children (approximately 28 per group) with BLLs spanning a range from approximately 5 to 10 µg/dL (Factor Litvak et al. 1996). In a population of 12- to 15-year-old children living near a lead smelter and a control group, urinary retinal binding protein (U-RBP) was found to be significantly associated with BLL in a stepwise regression. When urinary RBP excretion was examined by BLL tertiles, significantly lower U-RBP was seen in the group with BLL < 8.64 µg/dL compared with BLLs 8.64–12.3 µg/dL.

Heme synthesis biomarkers

Roels et al. (1987) studied the relations of PbB to heme synthesis biomarkers and reported no evident threshold for inhibition of aminolevulinic acid dehydratase synthesis at PbB as low as 8–10 µg/dL, while the threshold for increasing erythrocyte protoporphyrin levels was evident in the range of 15–20 µg/dL, consistent with other studies, including two meeting criteria for inclusion in this report (Rabinowitz et al. 1986; Hammond et al. 1985).

Discussion

Question 1: Does available evidence support an inverse association between children's blood lead levels <10 µg/dL and children's health?

The weight of available evidence, both indirect and direct, clearly favors an inverse association between these BLLs and cognitive function among children. The indirect evidence comes from the great majority of studies that have examined BLLs in relation to standardized measures of overall cognitive function; these studies reveal an inverse relationship and no trend toward weaker associations in populations with lower BLL distributions. More direct evidence of such an association comes from a recent analysis of data from a cohort designed from the start to study the relation of blood lead to child development (Canfield et al. 2003). This study demonstrated that the inverse relation between BLL and cognitive function exists and is stronger at BLLs <10 µg/dL compared with higher BLLs; it also does not show a threshold within the range of routinely measured BLLs below which no association was present. A recent letter to the editor described a reanalysis of the Boston cohort data (Bellinger et al. 2003) with findings consistent with Canfield et al (2003). Several recent analyses of data from the NHANES III and other populations also provide direct evidence of associations that imply adverse impacts of lead on indicators of children's neurocognitive development, stature, head circumference, dental caries, and sexual maturation in girls, occurring at measured BLLs <10 µg/dL. Though the number of studies providing direct evidence of associations at BLLs <10 µg/dL is limited and most are cross-sectional, they provide supporting evidence of an association in the context of the much larger number of studies that relate slightly higher levels of lead in blood to impairments of children's health.

Question 2: Are the observed associations likely to be causal?

Though the weight of evidence favors an association between children's BLLs <10 µg/dL and health and, indeed, suggests that such relationships become steeper as BLLs decrease, the WG considered a number of concerns that must be addressed in judging whether such associations are likely to be causal. The work group concluded that collectively, these concerns and limitations of the available evidence preclude definitive conclusions about causation and leave considerable uncertainty concerning the magnitude and form of causal relations that may underlie these associations. At the same time, available evidence does not refute the interpretation that these associations are, at least in part, causal. These issues are discussed individually in the following text, followed by overall conclusions.

Biologic Plausibility

Evidence from experimental animal and in vitro studies, which are not subject to confounding influences of concern in human observational studies, can establish causation and identify mechanisms that might be operative in humans assuming a suitable animal model. Thus, evidence from experimental animal and in vitro studies can help to assess potential dose–response relationships and thresholds within the context of any uncertainty added due to interspecies extrapolation. Therefore, an important consideration in judging whether associations between BLLs <10 µg/dL and health outcomes are likely to represent causal relationships is whether such relationships are biologically plausible on the basis of experimental animal and in vitro studies. These studies can also help to assess potential dose–response relationships and thresholds, but extrapolation from in vitro and animal models to human health risk adds additional uncertainty.

Lead is the most extensively studied environmental neurotoxicant. Animal and in vitro studies have provided abundant information concerning biochemical and physiologic changes caused by lead. Along with clinical and epidemiologic data, this evidence has clearly established that lead is toxic to the developing and mature nervous system. These data have been extensively reviewed elsewhere (USEPA 1986; ATSDR 1999; WHO 1995; Davis et al. 1990) and are not exhaustively reviewed here. Rather, this discussion highlights evidence concerning potential mechanisms of lead toxicity and data from animal studies that are relevant to the biologic plausibility of the toxicity of lead, especially to the developing nervous systems of children exposed at BLLs <10 µg/dL.

Although the precise mechanisms of action and their relative importance in different manifestations of lead toxicity are not known definitively, in vitro studies demonstrate that lead can interfere with fundamental biochemical processes. At the most basic level, many of the proposed mechanisms of lead toxicity involve binding to proteins and/or interference with calcium dependent processes (Goldstein 1993).

For some of the adverse health effects of lead (e.g., anemia), the lead-associated biochemical changes that contribute to the effect in humans are well understood. Lead interferes with heme synthesis in part by binding to sulfhydryl groups in the enzyme amino levulinic acid dehydratase (ALAD) (ATSDR 1999), which is especially sensitive to inhibition by lead (less than 0.5 micromoles per liter in vitro) (Kusell et al. 1978; Dresner et al. 1982). This inhibition causes delta amino levulinic acid, a potential neurotoxic agent, to accumulate. Lead also inhibits ferrochelatase, an enzyme catalyzing the incorporation of iron into protoporphyrin to form heme. This inhibition also may involve lead binding to protein sulfhydryl groups.

Although anemia and accumulation of protoporphyrin IX in erythrocytes are the most obvious consequence of impaired heme synthesis, this pathway could play a role in lead-related impairment of cellular function throughout the body (USEPA

14

1986). By interfering with heme synthesis and perhaps by inducing enzymes that inactivate heme, lead can decrease the levels of heme in body tissues (Fowler et al. 1980). A reduction in the body heme pool may impair heme-dependent biochemical processes, such as cellular respiration, energy production, and the function of the cytochrome p-450 monooxygenase system involved in detoxification of xenobiotics and in transformation of endogenous compounds such as vitamin D precursors (USEPA 1986).

For other more complex health effects of lead, such as impaired neurocognitive development and behavioral change, a number of plausible mechanisms have been demonstrated in animal and in vitro systems. Lead's impact on one or more biochemical systems needed for normal brain development and function could account for the neurobehavioral effects observed at low levels of exposure. Especially sensitive to lead in vitro is the activation of protein kinase C (PKC), a calcium dependent enzyme. Lead binds more avidly to PKC than its physiologic ligand, calcium, causing activation at picomolar concentrations in vitro. (Markovac and Goldstein 1988). The interactions between lead exposure and PKC activity in the brain are complex; chronic lead exposure may reduce activity of PKC associated with cell membranes while increasing cytosolic PKC activity. Lead effects on PKC activity have been proposed to mediate potential impacts of lead on cell growth and differentiation, including that of neural cells (Deng and Poretz 2002), the blood-brain barrier, and long-term potentiation (a process related to memory) (Hussain et al. 2000). Lead also interferes with calcium-dependent control of neurotransmitter release at presynaptic nerve terminals; it may thereby interfere with signaling between neurons and possibly with development of neural networks. In both animal and in vitro studies, lead has been demonstrated to interfere with neurotransmitter systems, including interfering with dopamine binding and the inhibition of N-methyl-D-aspartate (NMDA) receptor activity.

The large body of evidence from animal studies of lead exposure and neurodevelopment supports a causal effect that is persistent following exposure early in life and that generally parallels human studies in terms of the domains of function that are impaired (WHO 1995). Concerning blood lead–effect relationships, direct cross-species comparisons of BLLs cannot be made (Davis et al. 1990), and most animal studies demonstrating lead-related developmental neurotoxicity involved doses that produced BLLs well above 10 µg/dL. However, available studies provide strong evidence of adverse effects in animals with BLLs near 10 µg/dL. It should be noted that BLLs cited in animal studies generally involve mean levels achieved in experimental groups with individual animals varying, sometimes substantially, around that mean.

Non-human primates experimentally exposed to lead early in life demonstrate dose related impairments in learning and behavior (Bushnell and Bowman 1979; Rice 1985; Levin and Boman 1986). One study, involving monkeys dosed during the first 200 days of life with 100 µg/kg/day lead or 50 µg/kg/day lead resulting

15

in average peak BLLs of 25 and 15 µg/dL respectively, showed deficits relative to control monkeys (dose=0µg/kg/day lead; average peak BLL=3 µg/dL) at age 3 years on "discrimination reversal" tasks (the animals are taught to respond to a cue and then the cue is changed and the ability to learn the new cue, with and without irrelevant cues, is measured). At the time of testing, mean BLLs in the exposed groups had fallen to 13 and 11 µg/dL, respectively. Both exposure groups showed deficits, but deficits in the lower exposed group were evident only with more complex tasks (e.g., including irrelevant cues) (Rice 1985). The same monkeys showed persistent impairments at 9 to 10 years of age (Gilbert and Rice 1987). Experimental studies in rats have demonstrated behavioral effects at mean BLLs of 10–20 µg/dL (Cory-Slechta et al. 1985; Brockel and Cory-Slechta 1998).

There is uncertainty about the relationship of the tissue or cellular levels of lead linked to physiologic changes in animal and in vitro studies to the corresponding human blood lead level required to produce such levels at target sites. Although most (90 to 99%) lead in whole blood is in red cells, plasma lead level likely better reflects lead transferred from bone stores and available for transfer to target tissues (Cake et al. 1996). Because red cells have limited capacity to accumulate lead, the relation of blood lead to plasma or serum lead is non-linear with serum lead increasing more rapidly at higher BLLs (Leggett 1993). In subjects with a mean BLL of 11.9 µg/dL, plasma lead levels ranged from 0.3 to 0.7% of whole BLLs (Hernandez-Avila et al. 1998). The relation of plasma serum levels in intact animals to tissue levels measured in in vitro models is probably more complex. It is also uncertain whether in vitro studies demonstrating possible mechanisms for low-level lead toxicity reflect mechanisms operative in the intact animal. For example, Zhao et al. (1998) found that lead interfered with PKC in choroid plexus endothelial cells in a dose dependent fashion over the concentration range of 0.1-10 micromolar. However, no effect on choroid plexus PKC activity was seen in an in vivo model.

Conclusions: The fundamental nature of biochemical and physiologic changes linked to lead in in vitro and experimental animal studies illustrates potential mechanisms for lead toxicity that might be operative in humans at very low exposure levels. Experimental animal studies support the biologic plausibility of adverse health effects of lead in children at BLLs near 10 µg/dL. However, definite conclusions concerning the relationship of health status of children and BLLs <10 µg/dL cannot be drawn from these studies because of limitations of extrapolating from in vitro systems to intact animals and from animals to humans and because of the limited amount of data available from studies of animals dosed to produce a range of BLLs less than 10 µg/dL. Data from primates, which can most readily be extrapolated to humans, are especially limited.

On the other hand, given the uncertainty in extrapolating across species, the fact that animal test systems cannot match the complexity of learning tasks faced by young children, and the relatively small relative difference in BLLs shown to be harmful in animals and those at issue in children, adverse health effects in children at BLLs <10 µg/dL are biologically plausible.

Blood Lead Measurement

The precision and accuracy of blood-lead measurements performed in an epidemiologic study impacts observed results. If BLLs are systematically over or underestimated, biases in estimated blood lead response relationships and/or no effect thresholds will result. All blood lead measurements involve some random error, which, if a true association between blood lead and health exists, will tend to bias estimates of the relation toward the null (i.e., no effect) value. The quality of blood lead measurements varies between laboratories, between different analytical technologies, and between different specimen collection techniques. In addition, laboratory performance for blood lead has improved markedly over the last three decades and continues to improve as new analytical technologies are developed. Each of these factors becomes important in assessing the quality of blood-lead measurements used in published studies. In this section, specimen collection and laboratory factors that can affect blood-lead precision and accuracy are considered.

The widespread industrial use and dispersal of lead, particularly during the last century, has ensured that it is a ubiquitous contaminant. Therefore, to prevent false-positive results, stringent procedures are necessary to reduce environmental contamination of blood collection devices and supplies. Consequently, venous blood collected using evacuated tubes and needles certified as "lead-free" is considered the most appropriate specimen for blood lead measurements (NCCLS 2001). However, collection of venous blood from pediatric subjects is sometimes difficult; thus, capillary blood from a finger puncture is used widely for screening purposes. Published studies have compared the quality of blood lead results for capillary and venous specimens drawn simultaneously (Schlenker et al. 1994; Schonfeld et al. 1994; Parsons et al. 1997). With stringent precautions, particularly rigorous hand washing, contamination errors can be held to <4% (Parsons et al. 1997). Therefore, although venous blood is preferable for epidemiologic studies of environmental lead exposure, use of capillary blood is acceptable if collected by staff specially trained in the technique using devices certified as "lead-free." Data should be provided showing an acceptably low rate of contamination errors and low mean bias in the capillary BLLs as collected using the study protocol.

Currently, three analytical approaches to blood lead measurement are used: atomic absorption spectrometry (AAS); anodic stripping voltammetry (ASV), and inductively coupled plasma mass spectrometry (ICP-MS). A thorough discussion of these analytical techniques is beyond the scope of this report; however, a comprehensive assessment has been published by the National Committee for Clinical Laboratory Standards (NCCLS 2001). Briefly, the older flame atomization AAS methods, which include MIBK-extraction and Delves cup, are less precise, with a detection limit near 5 µg/dL for Delves cup (Parsons and Slavin. 1993). Thus, they are not well suited for examining relationships between BLLs <10 µg/dL and health. The electrothermal atomization techniques based on the graphite furnace (ETAAS) are more precise and more sensitive and, therefore, have better detection limits, typically around 1.0 µg/dL. A direct comparison between ASV and ETAAS techniques (Bannon et al. 2001) shows that the latter has better precision and better accuracy. Nonetheless, when operated in experienced hands and with a stringent quality assurance/quality control (QA/QC) program that includes calibration standards traceable to the mole via isotope dilution mass spectrometry (ID-MS), ASV can deliver blood-lead measurements with accuracy and precision sufficient to examine health effects at BLLs <10 µg/dL (Roda et al. 1988).

In order to assess the accuracy and precision of blood-lead measurements made for research purposes, investigators should provide information on the laboratory's performance in measuring external quality control samples and on the between-run standard deviation for routine quality control samples that span the relevant blood lead range for a given study.

Conclusions: The key considerations relevant to judging the accuracy and precision of blood lead measurements in published studies include the type and quality of blood specimen collected, analytical methodology used by the laboratory, and internal and external QA/QC procedures in place. For the purpose of studying the relationship between BLLs <10 µg/dL and health endpoints venous samples are preferred and capillary samples are acceptable with evidence of a rigorous protocol to control contamination errors. Acceptable analytic methods include electrothermal AAS, ASV, and ICP-MS. Information on laboratory performance (i.e., accuracy and precision) from external and internal quality control data should be provided.

To be included in this review, studies were required to have employed suitable measurement methods. In addition, venous samples were used for most postnatal blood lead measurements in the relevant cohort studies (Table 6) and others cited in this report. Given this and the blood lead quality control procedures reported in the most informative studies, it is highly unlikely that systematic errors in measurement in the relevant studies were sufficient to bias the observed blood lead distributions enough that associations observed <10 µg/dL were attributable to BLLs above that threshold. Random variation in BLLs and random error in BLL measurement would make it difficult to collect sufficient data to identify a threshold, if one were to exist.

Blood Lead Age Trend, Tracking, and Inference Concerning Blood Lead — Effect Relations in Children

Age-related changes in children's BLLs and within-child correlation of blood lead measured at different ages may influence observed associations between BLL and health at a given age. In addition, the biologic impact of lead in children is likely determined not only to the BLL measured at any one time but, also, the ages at which a given level occurs and the duration of exposure.

Under most exposure scenarios, children's BLLs show a characteristic age trend. A newborn's BLL will largely reflect the BLL of its mother. Because adult women tend to have lower BLLs than young children, umbilical cord BLLs are generally lower than BLLs during childhood. During the latter half of the first year of life, however, children's BLLs begin to increase as the infant becomes more active, mobile, and exposed to ambient lead. The onset of ambulation during this period is likely to be important, as are play patterns that bring the child into contact with environmental media such as lead-contaminated dust and soils. Other factors that affect exposure include the increased hand-to-mouth activity of children, including the practice of eating "in place," i.e., in play areas. Physiologic factors, such as more efficient absorption of ingested lead in children compared with adults, and their greater food and air intake on a body weight basis might also contribute to the early postnatal rise in BLL.

The mean BLL within a study sample generally peaks during 18 to 36 months of age, and slowly declines over the next few years. This blood lead profile is seen among economically disadvantaged urban minority children (Dietrich et al. 2001) and among children living near lead smelters (Figure 4) (Tong et al. 1996). In cohorts with extremely high exposures, the blood lead decline might be very gradual (e.g., Wasserman et al. 1997). In the Cincinnati Study, the same general profile was evident in each of four strata defined by average lifetime BLL, suggesting that it is, to some extent, independent of the overall level of exposure. This blood lead profile has not been observed in all study cohorts, however. In the Boston Study, for example, mean BLL varied minimally, from 6.2 to 7.6 µg/dL, from birth through 5 years of age (Rabinowitz et al. 1984; Bellinger et al. 1991).

One implication of the typical profile is that maximum level is often associated with age, constituting an obstacle to an effort to identify age-specific vulnerability to lead toxicity. Compounding this challenge is that, under many exposure scenarios (particularly those involving higher exposures), intra-individual stability of BLL tends to be substantial. That is, BLL tends to "track," so that if, at time 1, child A has a higher BLL than child B, child A is likely to have a higher BLL than child B at time 2 as well. Thus, children's rank ordering tends to be similar over time even though, in absolute value, BLL rises and falls over the course of childhood. Again, however, the degree of intra-individual stability varies from cohort to cohort. In the Boston Study cohort, for instance, the extent was limited; this stability is likely attributable to the generally low BLLs of the study population (Rabinowitz et al. 1984).

A BLL measured after 36 months of age will, on average, be lower than the BLL that would have been measured if a child's blood been sampled sometime during the 18 to 36 month period. Suppose, however, that the critical period with regard to producing an adverse health outcome is the 18 to 36 month period, and that, in a study conducted post-36 months, an inverse association is noted between concurrent BLL and a health endpoint. If the concurrent BLL is the only index of lead exposure history available, basing a dose–effect assessment on it will, to the extent that the natural history of BLLs in the study cohort follows the canonical form illustrated above, result in an underestimate of the BLL responsible for any adverse health effects noted at the time of or subsequent to blood sampling. In other words, one will conclude that adverse health effects occur at lower BLLs than is the case. For instance, assume that the inverse association shown in Figure 5 holds between IQ and concurrent blood lead in a cross-sectional study of 6 year olds.

If, however, the BLL of each child was, on average, 5 µg/dL greater at age 2 than at age 6, and age 2 is the time of greatest toxicologic significance (i.e., it is age at which lead exposure produced the IQ deficit observed at age 6), then the dose–effect relationship that underlies the association seen at age 6 would be more accurately described as in Figure 6.

This dataset would thus not be informative with respect to the functional form of the dose–effect relationship at levels below 10 µg/dL insofar as (hypothetically) all children had a BLL greater than 10 µg/dL at age 2.

Other uncertainties apply to interpreting blood lead–health associations (or lack of associations) observed at any point in time. First, the relation of age to vulnerability to lead toxicity is not well understood. Is BLL during the age period 18 to 36 months more toxicologically critical than a measure of cumulative lifetime exposure, such as the area under the curve or some other exposure index? Also, it is possible that the critical age varies with dose, health endpoint, or sociodemographic factors. Available studies do not provide consistent answers to these questions. For example, in the Boston cohort, blood lead at age 24 months was most strongly related to IQ at age 10 years (Bellinger et al. 1992), whereas in the Port Pirie Cohort, the lifetime average BLL through age 5 years was most predictive of IQ at age 11 to 13 years.

If 18 to 36 months is the critical age of exposure, theoretically, it is possible to "adjust" an observed blood lead distribution measured at age 6 by some function to reflect the downward trend in BLL with age and estimate the blood lead distribution at a different age, (e.g., age 2 years). However, a "one size fits all" adjustment likely is not appropriate for all children. Moreover, the appropriate adjustment is likely to be study-site-specific (i.e., depend on the key exposure sources and pathways of a particular study cohort).

It would be possible to get a general sense of how accurately past peak exposure can be estimated for children in cross-sectional studies by using data collected in prospective studies in which blood lead was measured frequently during the period spanning birth to school-age. Examining the distribution of the differences between BLLs measured at ages 18 to 36 months and at age 6 would suggest the amount of exposure misclassification that would result from applying a constant adjustment factor.

Conclusions: Because of age trends in blood lead and the tendency of BLLs to "track" within individual children, inferences drawn from cross-sectional associations between blood lead and health at a given age should be interpreted cautiously because of the influence of likely higher BLLs occurring earlier in life. It may be possible to apply data on age trends and within-subject correlation of blood lead to estimate, from an observed blood lead–health association, the approximate relation to BLLs at an earlier age. However, because age trends and the extent of "tracking" of blood lead levels vary from one population to another, it is not possible to estimate with confidence the distribution of blood lead levels earlier in life for any given population whose blood lead levels were only measured at one point in time. If the only relevant studies available are based on cross-sectional data (e.g., data from NHANES III), age trends and "tracking" of BLLs would represent a substantial challenge to inferring a causal link between BLLs <10 μg/dL and adverse health impacts. However, recently published results from two cohort studies (Canfield et al. 2003; Bellinger et al. 2003) showed inverse associations between BLLs measured early in life (6 to 24 months and 24 months, etc.) and IQ measured at older ages among children whose measured BLLs did not exceed 10 μg/dL. Therefore, associations observed in cross-sectional studies cited in this report likely do not exclusively result from the impact of higher BLLs experienced earlier in life.

Quality of Neurobehavioral Assessments

As with blood lead (exposure) measurements, the accuracy, precision, and consistency of neurobehavioral assessments can influence observed blood lead-outcome relations. In order to judge whether the data from a study should be considered in characterizing the functional form of the dose–effect relationship at BLLs <10 μg/dL, one would like to have access to the following information about the conduct of the neurobehavioral assessments:

- Assurance that examiners were blinded to all aspects of children's lead exposure histories.

- The assessment setting. Assessments can be standardized when carried out in a hospital, neighborhood health center, or community center, but may be difficult to standardize in a participant's home.

- Essentials of the process by which an examiner was trained, including the criterion used to certify an examiner (e.g., percent agreement on an item-by-item basis with some gold standard, average difference in scores assigned compared with gold standard, correlation with gold standard in terms of scores assigned).

- The plan implemented for supervision of test administration over the course of data collection (e.g., periodic observation of test sessions, live or by videotape).

- The plan implemented for supervision of test scoring over the course of data collection (e.g., double scoring of a sample of protocols).

- The number of neurobehavioral examiners used over the course of data collection.

- If more than one assessor was used, whether the data analysis plan included evaluation of an "assessor" effect (i.e., as a main effect and as a modifier of lead's association with endpoints).

While some have argued that neurobehavioral examiners should have professional qualifications (e.g., Kaufman 2001 cites the need for a clinician with graduate-level training in psychometrics, neuropsychology, etc.), the Practice Committee of the American Academy of Clinical Neuropsychology supports the widespread practice of using non-doctoral level personnel, with appropriate training and supervision by a doctoral-level psychologist, in the administration and scoring of clinical neuropsychological evaluations (Brandt et al. 1999).

Assuming examiners are blinded regarding BLLs, most problems with quality of neurobehavioral assessment would be expected to mask or underestimate true associations rather than create spurious ones. It is possible, for example, that use of non-professional examiners might introduce noise into the data, masking an association between toxicant exposure and performance. In one study of methylmercury (MeHg) exposure (Grandjean et al. 1997), MeHg was inversely associated with children's scores on the Similarities subtest of the WISC-III among children tested by the supervising doctoral-level study examiner. Assuming that blinding was preserved, use of non-professional examiners likely would not introduce a positive bias in effect estimates.

Measurement quality problems causing bias of associations away from the null, without loss of blinding, are theoretically possible. For example, if one examiner consistently yields lower scores than another and that examiner, without knowledge of BLLs, is assigned to assessments of a segment of the study population at higher risk for lead exposure, a spurious inverse association could be created between lead level and neuropsychological test scores.

Conclusions: The key considerations in judging the quality of neurobehavioral assessments in the research setting are the blinding of examiners to lead-exposure history, the training and supervision of examiners, and the setting for examinations. If examiners are truly blinded, other data quality problems generally will bias estimated relationships between blood lead and outcomes toward the null. Therefore, given that examiners were blinded to BLLs in cohort studies demonstrating associations (Table 6) and the NHANES III survey, errors in measurement of neuropsychological function likely did not contribute to observed associations with BLLs <10 µg/dL.

Potential Confounding Factors

Social Factors

Socioeconomic factors influence both lead exposure and many health outcomes, including intellectual development, growth, and a number of chronic conditions, creating the potential for social factors to confound associations between children's lead exposure and health in observational studies. Because cognitive function as reflected in measured intelligence is strongly associated with socioeconomic status (SES) and because cognitive function in children is the most studied health endpoint in studies of lead-exposed children, this discussion is focused on possible SES confounding of associations between BLL and measured intelligence. The potential for reported subtle effects of lead on IQ and related measures of intellect to be attributable to confounding by socioeconomic factors warrants serious consideration (Bellinger et al. 1989). Key relations required for confounding to occur are almost certainly present—SES has been shown to be related to BLLs, presumably because the neighborhoods and homes in which families of lower income reside are associated with higher levels of lead in soil and residences. Socioeconomic status also is clearly related to measures of intelligence, whether through parental stimulation, nutrition, or resources available in the home. With an inverse relationship between socioeconomic factors and lead levels (i.e., higher SES predictive of lower lead levels) and a positive relationship between socioeconomic factors and measures of intelligence (higher SES predictive of higher intelligence test scores), failure to adjust for the confounding effect of socioeconomic factors will result in confounding that overstates the harmful effect of lead on IQ because the socioeconomic effect will be mixed with any true effect of lead exposure. In addition, confounding by social factors may be a concern for some other lead-associated health measures with social gradients such as height (Silventoinen 2003).

Data presented from most of the key studies included in this review strongly suggest that substantial confounding by socioeconomic factors occurs. Even with adjustment for crude measures (e.g., parental education and household income) (Lanphear et al. 2000), the apparent lead effect on cognitive function is greatly reduced. Such a pattern in which adjustment for a crude proxy results in a substantial decrement in the magnitude of association would suggest that "residual confounding" may be present in the adjusted estimate of effect. If residual confounding is indeed present, then tighter control for confounding with more refined measures of the social environment may further attenuate or eliminate the apparent effect (Savitz et al. 1989).

The following factors complicate this scenario:

1. Socioeconomic status is a very elusive construct to fully capture; it is far more complex than is reflected in parental education or income. Socioeconomic status includes many aspects of economic means and associated lifestyle,

so that adjustment for operational measures, such as education or income, will always be incomplete. Adjustment for an imperfect proxy measure of a confounder results in residual confounding (Greenland et al. 1985; Savitz et al. 1989).

2. Long-term lead exposure is imperfectly reflected in a current blood lead measure or to some extent, even from a series of blood lead measures (see Blood Lead Tracking) (Bellinger et al. 1989). Whatever physiologic effect lead might produce, available evidence suggests that the impact is chronic and cumulative. Beyond what is reflected in a blood-lead measure, SES may be indicative of historical exposure; thus, the observed effect of socioeconomic status would partly reflect an effect of lead exposure above and beyond the blood-lead measure.

The nature and magnitude of these associations is less clear when focusing on BLLs <10 µg/dL. Measures of social advantage, including income and parental education, are associated with BLLs <10 µg/dL (e.g., Lanphear et al. 2000). However, the relative importance of different aspects of socioeconomic status and the pathways by which they affect lead exposure are not entirely clear. The association between lower income and deterioration of paint in older housing contributes to variation in BLLs, even BLLs <10 µg/dL. The increase in geometric mean blood lead associated with living in an older home is greater for children from low-income families than for those from middle income families (Pirkle et al. 1998). Nonetheless, with the elimination of lead in gasoline and the continued decline in the proportion of homes with leaded paint (Jacobs et al. 2002), the relative importance of lead exposure sources possibly is changing. It is also possible that the association of social factors with lead exposure is different for populations with BLLs <10 µg/dL than for those above that level.

Several strategies have been applied to address the role of socioeconomic factors and isolate a non-specific effect of socioeconomic factors on IQ from an effect of lead exposure. First, populations can be sought or even constructed in which blood lead is not closely associated with SES as demonstrated most clearly in the Boston cohort (Bellinger et al. 1987). In that population, all in a relatively low blood lead range for that time and the great majority of relatively advantaged SES, there was a weak positive gradient between socioeconomic status and lead. The Kosovo cohort (Wasserman et al. 1997) also departed from the usual trend in that the more SES advantaged of the two communities studied was the site of a lead smelter. As a result, adjustment for social and other covariates actually strengthened the inverse relation of blood lead to IQ in that population.

Second, improved measures of socioeconomic factors have been applied to better control for non-specific effects. That is, by refining and decomposing the construct of socioeconomic status, it is possible to adjust more fully for the confounding dimensions such as nutrition, parental stimulation, attitudes towards achievement, etc., and not adjust for the aspects that primarily serve as a proxy for lead exposure, such as age of housing and neighborhood. One example among published research of refining and decomposing the construct of socioeconomic status has been the use of the HOME scales to adjust for stimulation provided by caregivers. Use of HOME scales has in some cases further attenuated but not eliminated apparent lead–IQ associations.

A third approach to examine the possibility of confounding of the blood lead–IQ relation at low levels would be to conduct a formal statistical assessment of the extent to which the strength of the observed association across studies varies in relation to control for relevant confounders, using meta-regression, as was applied by Schwartz (1994). This approach could be refined to assess possible residual confounding. One challenge in performing such an analysis using published summary data is the difficulty in operationalizing measures of the tightness of SES adjustment while controlling for other aspects of study design that might influence blood lead–IQ slopes. An alternative approach is discussed later in this report (see Research Needs).

Conclusions: On the basis of available evidence the observed associations between blood lead below 10 µg/dL and cognitive function likely do not entirely result from confounding. This conclusion is supported by the following evidence:
- The studies showing the strongest relationship (Canfield et al. 2003; Bellinger et al. 2003) at low levels employed the HOME scale for adjustment, which is the best available measure for assessing the impact of the home environment on child development.
- Two cohorts, Kosovo and Boston, in which strong associations were found between blood lead and IQ, were characterized by a direct, rather than inverse, correlation of blood lead with social advantage.
- Associations of children's blood lead close to 10 µg/dL and intelligence have been seen in diverse geographic and social settings.
- Animal data have demonstrated effects of lead at BLLs near 10 µg/dL.

On the other hand, the ability to detect confounding by omitted covariates by comparisons across studies is limited because, for most covariates of potential interest, the number of relevant studies in one group being compared is limited. In other words, for a given covariate, either few studies included it (e.g., postnatal ETS exposure) or few excluded it (e.g., SES). At this point, the case for residual confounding by social environment is speculative, but available studies relating blood lead to cognitive function in children cannot entirely exclude the possibility that observed associations are at least partly influenced by it. Such a possibility does increase uncertainty about the actual strength and shape of blood lead relationships at BLLs less than 10 µg/dL.

Iron Status

Nutritional factors, such as iron and zinc intake, that might be correlated with lead uptake and might influence children's health, could confound associations between BLLs and health from observational studies. The potential for iron deficiency to confound the association between blood lead and neurodevelopmental status has been of most concern and is the focus of this discussion. The likelihood of such bias is dependent upon the extent to which iron status was controlled in a given study and the prevalence of iron deficiency in a study population. Iron deficiency may impair neurodevelopment in a manner similar to low-level lead exposure and the populations at increased risk for iron deficiency and lead toxicity may overlap (Lozoff et al. 1991; Wasserman et al. 1999). However, the association between iron deficiency and blood lead is not consistent across populations (CDC 2002). Therefore, the potential for iron to confound an association of blood lead with neurodevelopmental status will vary across populations, depending on both the prevalence of iron deficiency and its association with blood lead level.

For research purposes adequate assessment of iron status entails determination of hemoglobin or hematocrit and at least two other measures of iron status. Generally accepted definitions of iron deficiency and iron deficiency anemia depend on age- and sex-specific normal ranges. The iron status measures most commonly used are mean corpuscular volume (MCV), free erythrocyte protoporphyrin (FEP) or zinc protoporphyrin (ZPP), transferrin saturation, ferritin, and, more recently, transferrin receptor. The standard for defining iron deficiency is values indicating iron deficiency on at least two of these measures and/or response to iron therapy with an increase in hemoglobin to at least 10 g/L. The utility of ferritin in young infants is under debate, making it important that functional measures, such as MCV or ZPP be obtained. There are limitations of each measure (e.g., ferritin goes up with infection, MCV is down in hemoglobinopathies, etc.).

Although iron deficiency with low hemoglobin has been associated with later impairment of cognitive function (Grantham-McGregor et al. 2001), it is not certain which measure(s) of iron status are most strongly related to neurodevelopmental outcomes. In studies of children with higher BLLs, controlling for hemoglobin is problematic because lead toxicity can reduce hemoglobin in the normal range or cause frank anemia. This is less of a concern in studies of children with BLLs <10 μg/dL, a range in which no meaningful impact on hemoglobin levels has been observed.

Conclusions: Measurement of iron deficiency has been absent or suboptimal in most of the studies reviewed. Two studies in which iron status was controlled for using transferrin saturation (Canfield et al. 2003) and serum ferritin (Lanphear et al. 2000) found strong inverse relationships between blood lead and cognitive function, whereas a third study that controlled for the presence of iron-deficiency anemia found the opposite (Wolf et al. 1994). Furthermore, iron-deficiency anemia is the measure of iron status most clearly linked to impaired cognitive function; therefore, it seems unlikely that the prevalence of iron deficiency anemia could be high enough in the populations showing the strongest inverse relations of blood lead to cognitive function (Canfield et al. 2003; Bellinger et al. 2003; Lanphear et al. 2000) to entirely explain these associations. In the NHANES III data used by Lanphear et al. (2000), the prevalence of iron deficiency ranged from 1% to 9%, depending on the age and sex group (CDC 2002). Finally, in Kosovo, following treatment of iron-deficient children with iron supplements, no association of earlier hemoglobin levels with IQ at age 4 (Wassserman et al. 1994) or age 7 (Wasserman et al. 1997) was found. Thus, iron deficiency likely does not completely explain the inverse associations between BLLs <10 μg/dL and cognitive function.

Tobacco

Blood lead levels in children have been associated with exposure to environmental tobacco smoke (assessed by caregiver report or by urinary cotinine levels) in both general population surveys (Stromberg et al. 2003; Mathee et al. 2002; Lanphear et al. 2000; Mannino et al. 2003) and in studies of children living near lead smelters (Willers et al. 1988; Baghurst et al. 1992; Baghurst et al. 1999). The explanation for this association is not entirely clear; possibilities include enhancement of lead uptake by environmental tobacco smoke (ETS), exposure to lead in ETS itself, and differences in cleaning practices or child supervision between households with and without smokers.

Maternal smoking during pregnancy has been associated with behavioral problems and impaired cognitive development in children; fetal hypoxia is one possible contributing mechanism (Habek et al. 2000). Evidence for an effect of prenatal or postnatal ETS exposure on neurodevelopment is less clear (Eskenazi et al. 1999). As with studies of lead and neurodevelopment, social factors may confound, at least in part, the association between maternal smoking and neurodevelopment (Baghurst et al. 1992). A child's prenatal exposure to maternal smoking or pre- or postnatal exposure to ETS could, if these are causally related to impaired neurodevelopment or other adverse health outcomes, confound the observed associations of lead and health. In addition, if a relationship between postnatal ETS and neurodevelopment is established, lead exposure could be a mediating factor.

Conclusions: Of the studies reviewed, most did not assess prenatal or postnatal ETS as a possible confounding factor. Those that assessed tobacco at all controlled for maternal smoking during pregnancy. However, the two exceptions, Lanphear et al. (2000) in which serum cotinine measurements were used to control for ETS and a study based on the Port Pirie cohort (Tong et al. 1996; Baghurst et al. 1992) which reported postnatal parental smoking, provide no evidence that confounding by tobacco exposure accounts for the associations observed between blood lead and adverse health effects. Limitations in available studies leave some uncertainty as to what contribution, if any, ETS might make to observed associations between BLL and health.

Causal Direction

Inference of causation from observational epidemiologic studies is sometimes complicated by the possibility that the health outcome under study could be a cause of the exposure or causally related to a third factor which itself is a cause of the exposure under study. Two factors that influence blood lead levels—mouthing behavior and calcium balance—are relevant to assessing causal direction in studies of the health effects of lead at low levels.

Mouthing behavior

An important pathway of lead uptake by young children is ingestion of lead-contaminated dust (Charney et al. 1980; Bornschein et al. 1985), presumably through mouthing of hands, surfaces, and objects on which the dust is deposited. Although mouthing behavior is difficult to measure, children with more reported mouthing behavior have higher BLLs in relation to environmental lead exposure (Lanphear et al. 1998; Bellinger et al. 1986; Baghurst et al. 1999). Pica (purposeful ingestion of non-food items) can be a consequence of impaired neurodevelopment and can predispose one to lead ingestion (Cohen et al. 1976; McElvaine et al. 1992; Shannon et al. 1996), but the relation of variation in "normal" age-appropriate mouthing behavior to neurodevelopment is uncertain. However, in groups of children, average measured or caregiver reported mouthing has been shown to diminish with age (Juberg et al. 2001; Tulve et al. 2002). Nonetheless, it is unclear whether, at the individual level, more frequent mouthing behavior is a marker (independent of its effect on lead ingestion) for delayed neurodevelopment. If such behavior is a marker, then an association between blood lead level and impaired neurodevelopment would result, and failure to adjust for mouthing behavior would result in an overestimate of the blood-lead effect. On the other hand, if measured mouthing behavior is associated with cumulative lead exposure above and beyond that reflected in measured BLLs, then controlling for mouthing behavior could amount to over control, underestimating the true effect of lead on neurodevelopmental measures.

Conclusions: At this point, no direct evidence supports reverse causation by mouthing behavior, and this hypothesis remains speculative. Arguing against this possibility, Tong et al. (1996) reported that an early measure of neurocognitive development, the Bailey MDI, was not predictive of later BLLs.

Calcium balance

Calcium balance changes in relation to growth during childhood and during the rapid expansion of bone mass during puberty and the pubertal growth spurt

(Bronner et al. 1998; van Coeverden et al. 2002; Bailey et al. 2000); estradiol may influence bone mineral deposition in pubertal girls (Cadogan et al. 1998). It is possible that effect of skeletal growth and puberty on calcium balance could cause lower BLLs (Thane et al. 2002), just as the opposite changes in calcium balance during menopause appear to cause an increase in blood lead (Hernandez-Avila et al. 2000; Garrido Latorre et al. 2003). It should be noted that the average age at menarche among U.S. adolescents dropped by approximately 2.5 months between the periods 1963-1970 and 1988-94 and that this trend was accounted for in part by a rising prevalence of obesity (Anderson et al. 2003). Average BLLs were likely falling substantially during this same period.

Conclusions: Because human studies linking blood lead at levels <10 µg/dL to delayed puberty and smaller stature are, with one exception, cross-sectional and evidence is limited on this topic, reverse causation via changes in calcium balance cannot be ruled out as accounting for at least some of the observed associations. While the parallel secular trends in decreasing age at menarche and decreasing BLLs could be explained in part to a causal effect of lead delaying age at menarche, it is also possible that other secular trends (e.g. increasing obesity rates) have caused the trend toward earlier menarche.

Overall Conclusions

Question 1: Does available evidence support an inverse association between children's blood lead levels <10 µg/dL and children's health?

Because of the large number of studies that have assessed cognitive function as an outcome, the review and conclusions by the WG primarily focus on this health domain. The consensus of the WG is that the overall weight of available evidence supports an inverse association between BLLs <10 µg/dL and the cognitive function of children. The evidence for such an association is bolstered by the consistency across both cross-sectional and longitudinal studies in varied settings with blood lead distributions overlapping 10 µg/dL and by the lack of any trend towards a weaker association in studies with lower population mean BLLs. More recent studies and analyses best suited to examining this association (Canfield et al. 2003; Bellinger et al. 2003) have added to, rather than refuted, evidence for such an association noted in prior CDC guidance (1991).

In reaching this conclusion, the WG is mindful of limitations in the available evidence base. Relatively few studies have directly examined the association between BLLs <10 µg/dL and health status among children and many of those that have are cross sectional studies in which data are unavailable on BLLs earlier in life and key covariates. The WG concluded that findings from numerous published studies relating BLL to cognitive function, while not limited to children with BLLs

<10 µg/dL, collectively were not consistent with a threshold for the BLL-cognitive function association at 10 µg/dL. This indirect evidence, however, is less persuasive than cohort studies and analyses that directly assess the relationship between BLL and health <10 µg/dL. These directly relevant studies analyzed data for children whose measured BLLs did not exceed 10 µg/dL (to the investigator's knowledge). Likely included in these analyses were some children who, because of random variation in BLL or age trends, did at some time have a BLL ≥10 µg/dL that was not measured. Such misclassification could produce an apparent inverse association between BLLs <10 µg/dL and health status even if a threshold existed at 10 µg/dL. Such misclassification, however, could not account for the observed BLL–IQ relation in the Canfield (2003) study, in which a steeper slope was observed at BLLs <5 µg/dL than at levels 5–10 µg/dL.

For health endpoints other than cognitive function, including other neurologic functions, stature, sexual maturation, and dental caries, available data are more limited and less replication of findings exists across studies. Nonetheless, the available data from these studies are consistent with associations between higher BLLs and poorer health indicators for values <10 µg/dL .

Question 2: Are the observed associations likely to be causal?

The work group concluded that, while available evidence does not permit a definitive causal interpretation of the observed associations between higher BLLs and adverse health indicators for values <10µg/dL, the weight of available evidence favors, and does not refute, the interpretation that these associations are, at least in part, causal. The WG also concluded that the limitations of the available evidence, including likely residual confounding by social environment, leave uncertainty about the absolute strength and shape of the causal relation at the population level. Even greater uncertainty attends the use of associations observed in the relevant population studies for interpretation of BLLs measured in individual children at a single point in time. Thus, the WG does not believe that the individual children can be classified as "lead poisoned," as the term is used in the clinical setting, on the basis of the associations observed in studies reviewed for this report. The basis of the overall WG conclusions is discussed below and is followed by a summary of the important limitations in the available evidence.

The WG explored other possible explanations (aside from causation) for these associations and concluded that none are likely to fully explain the observed data. The context of evidence from animal, in vitro, and human studies of adult populations, also supports the consensus of the WG conclusion that the observed associations most likely represent, at least in part, causal adverse impacts of lead on children's cognitive function at BLLs <10 µg/dL.

The greatest source of uncertainty in interpreting the relationship between BLLs <10 µg/dL and cognitive function is the potential for residual confounding by social factors. The conditions for residual confounding appear to be present: BLLs are strongly influenced by SES, SES is clearly related to measured cognitive function, and social factors that could influence BLL and cognitive function are difficult to measure precisely. Other sources of potential bias are, individually, less concerning than social confounding, but collectively they add to the overall uncertainty about the absolute strength and shape of the relation of BLL to impaired cognitive function. These include, random error in blood lead measurement and in a single BLL as a measure of chronic exposure, possible influence of factors that have not been fully addressed in published studies, including blood lead tracking and age trend, which limits cross-sectional studies in particular, tobacco smoke exposure, iron deficiency, and mouthing behavior. Error in measuring lead exposure would bias observed associations towards the null, while failure to adjust for the other factors noted would most likely bias observed associations away from the null.

The recently reported trend of asymptotically increasing slopes of lead-associated decrements in cognitive test scores at lower BLLs (Bellinger et al. 2003; Canfield et al. 2003; Lanphear et al. 2000) would be expected if residual confounding were operative as illustrated in Figure 7. The graph on the left depicts a comparison of two groups of children who live in a high exposure setting. They differ, on average, with respect to aspects of the home and social environment that are not captured in measured covariates. This results in one group ingesting and absorbing twice as much lead and having, after adjustment for measured covariates, a mean IQ 1 point lower than the children raised in a more favorable environment. Assuming a roughly linear relation of lead intake to blood lead, the result is that one group has a mean blood lead twice as high, corresponding to a 10 µg/dL difference in blood lead and an estimated blood lead-IQ slope attributable to residual confounding of 0.1 IQ points per µg/dL. The figure on the right depicts the same hypothetical two populations living in a low exposure setting. The same imperfectly measured differences in social environment contribute to the equivalent covariate-adjusted difference in mean IQ, but in this case, although one group ingests twice as much contaminated dust as before, lower levels of lead contamination result in the two children having a blood lead difference of only 1 µg/dL in blood lead level. The result is an estimated blood lead-IQ slope attributable to residual confounding of 1.0 IQ points per µg/dL. In addition, a convincing and directly relevant biologic mechanism for such a dose response relation has yet to be demonstrated. Though this hypothetical example cannot demonstrate that residual confounding underlies the steep blood lead-IQ slopes observed at low levels, it does support the need for caution in interpreting the absolute value of the estimated effect sizes.

The available data for these other health endpoints, taken mostly from cross-sectional studies, are more limited and firm conclusions concerning causation cannot be made at this time.

Research Needs

Resolving residual confounding through observational studies

It may be somewhat easier to identify study populations with BLLs <10 µg/dL in which socioeconomic factors are not associated with exposure as compared to populations with more widely varying blood lead levels within many low SES children, but few high SES children, may have blood lead levels above 20 or 30 µg/dL. Configuring a cohort similar to the one in Boston or assembling one from the pieces of others already studied could be helpful in isolating socioeconomic and lead effects from one another. Another formal statistical approach that could be applied to pooled data across multiple studies is the application of a hierarchical modeling approach as proposed by Schwartz et al. (2003, in press).

Controlled intervention trials

While experimental designs can establish causation with greater confidence than observational studies, intentionally exposing some children to higher BLLs in a randomized controlled design would be unethical. However, randomized trials in which interventions are tested for their ability to reduce BLLs <10 µg/dL or prevent their increase provide an opportunity to support or refute a causal relationship between BLLs <10 µg/dL and adverse health outcomes. Studies testing such interventions should measure covariates relevant to assessing health effects, allowing a test of the causal hypothesis should they be successful at sufficiently reducing BLLs.

Animal and in vitro studies to explore mechanisms and dose-response relations

While the overall evidence from animal or in vitro models supports the biologic plausibility of adverse effects of lead at BLLs <10 µg/dL, the WG is unaware of directly relevant animal or in vitro studies that demonstrate a steeper slope for adverse effects of lead exposure at lower BLLs than observed at higher levels. Demonstrating such a relationship in experimental studies and identifying possible mechanisms would increase confidence in a causal interpretation of the observed blood lead-response relationship in studies such as Canfield et al. (2003).

Figure 1. Expected variation in regression slopes given hypothetical

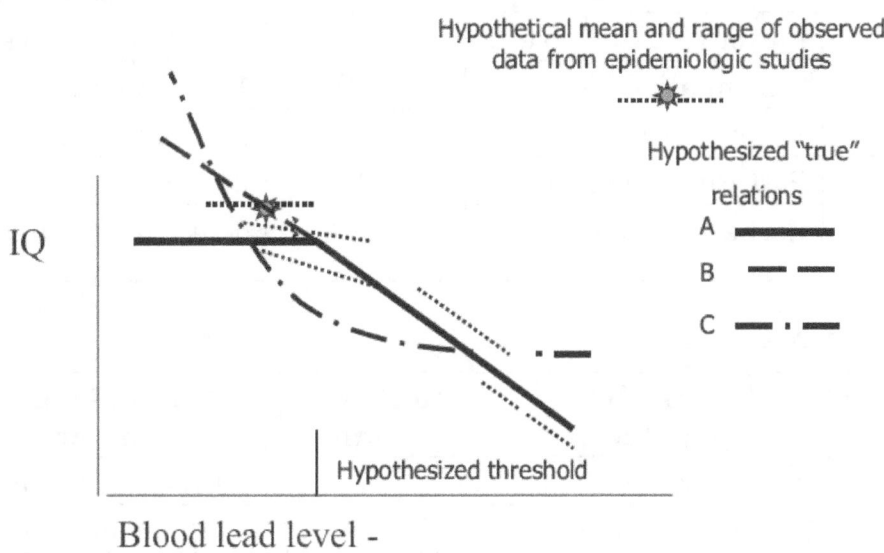

Notes: Figures 2 and 3

Selected estimates of change in outcome (Full Scale IQ or McCarthy General Cognitive Index (GCI) derived from regression coefficients and listed in Table 2 and the corresponding mean BLLs of the study population are displayed in Figures 2 and 3. Figure 2 contains results from studies where the BLLs were measured at ages ≤ 2 years, and outcome measures were measured at ages ≥ 4 years. Figure 3 contains the results when both the BLLs and outcome measures were measured at ages ≥ 4 years. Both the crude (open dot) and adjusted (solid dot) coefficients are displayed in the figures when both are available. (The Kosovo and European Multicenter Study papers did not provide the crude coefficient). Although multiple models for a single study population may have been fit to results from differing ages within the defined age categories, only the regression coefficients for the highest age at which blood lead was measured (per study population) are included in the figures. (The highest outcome measure age was used as a tiebreaker when necessary.) Also, when models for both a concurrent blood lead measure and a lifetime average blood lead measure existed for the highest age at which blood lead was measured (Port Pirie and Rochester), the concurrent results were included. For the study that provided multiple models for the same highest-age blood lead versus outcome measure (Lavrion, Greece), the results from the model that included the most covariates were included. Any studies not providing both a regression coefficient and blood lead mean were excluded. Three-letter abbreviations for each study population, defined in the legends below, were used on the plots.

Legend for Figure 2

Abbreviation	Study population	Reference number	Blood lead and outcome ages
Bos	Boston, Massachusetts	7	24 months / 10 years
Cin	Cincinnati, Ohio	13	15-24 months / 6.5 years
Kos	Kosovo, Serbia	37	24 months / 4 years
Por	Port Pirie, Australia	23	24 months / 4 years
Roc	Rochester, New York	11	6-24 months / 5 years

Legend for Figure 3

Abbreviation	Study population	Reference number	Blood lead and outcome ages
Bos	Boston, Massachusetts	7	10 years / 10 years
Cin	Cincinnati, Ohio	13	51-60 months / 6.5 years
Eur	European Multicenter Study	39	6-11 years / 6-11 years
Kar	Karachi, Pakistan	28	6-8 years / 6-8 years
Kos	Kosovo, Serbia	37	48 months / 4 years
Lav	Lavrion, Greece	20	primary school / primary school
Por	Port Pirie, Australia	35	11-13 years / 11-13 years
Roc	Rochester, New York	11	5 years / 5 years

Figure 2. Cognitive Function Regression Coefficients for Blood Lead Age ≤2 years and Outcome Age ≥4 years.

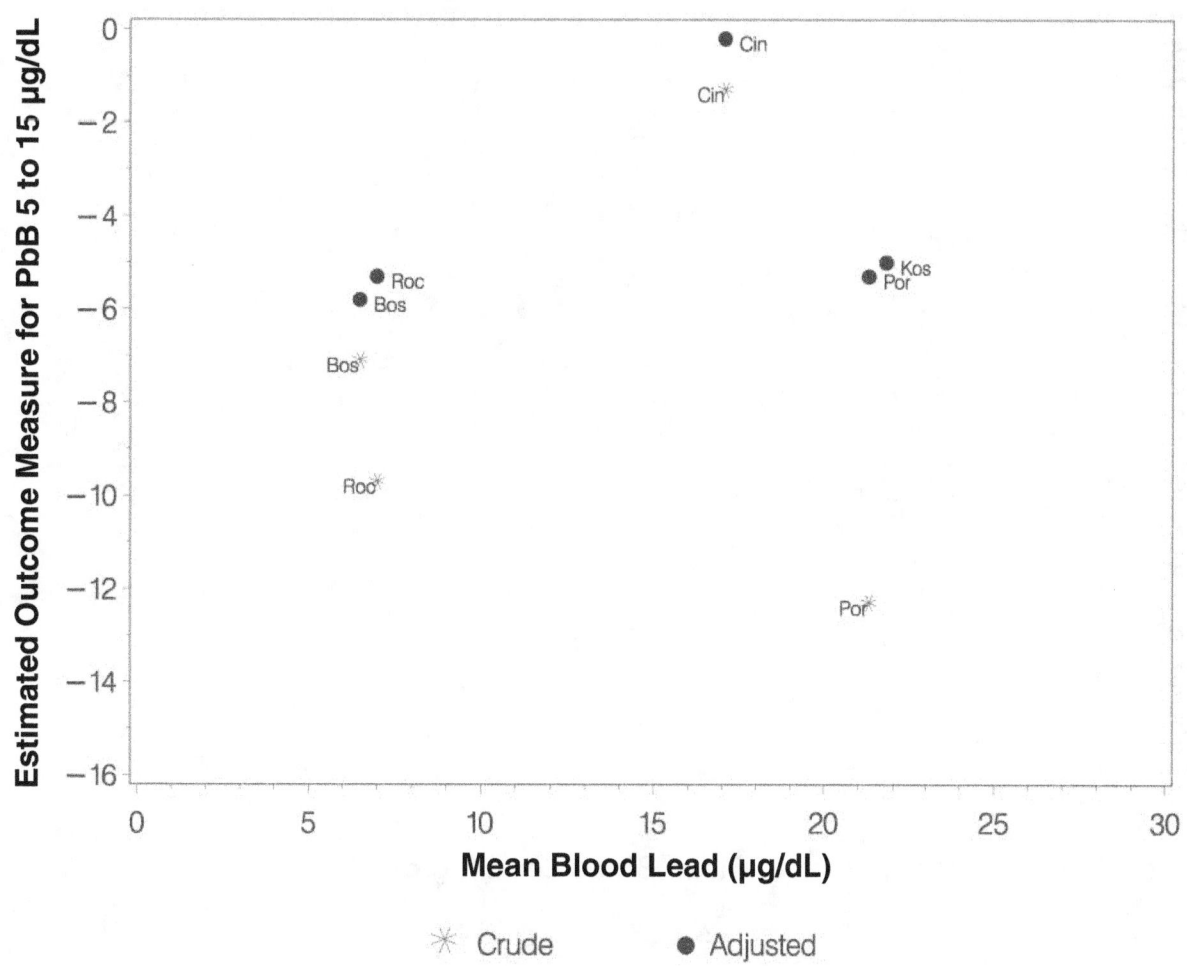

Figure 3. Cognitive Function Regression Coefficients for Blood Lead Age ≥4 years and Outcome Age ≥4 years.

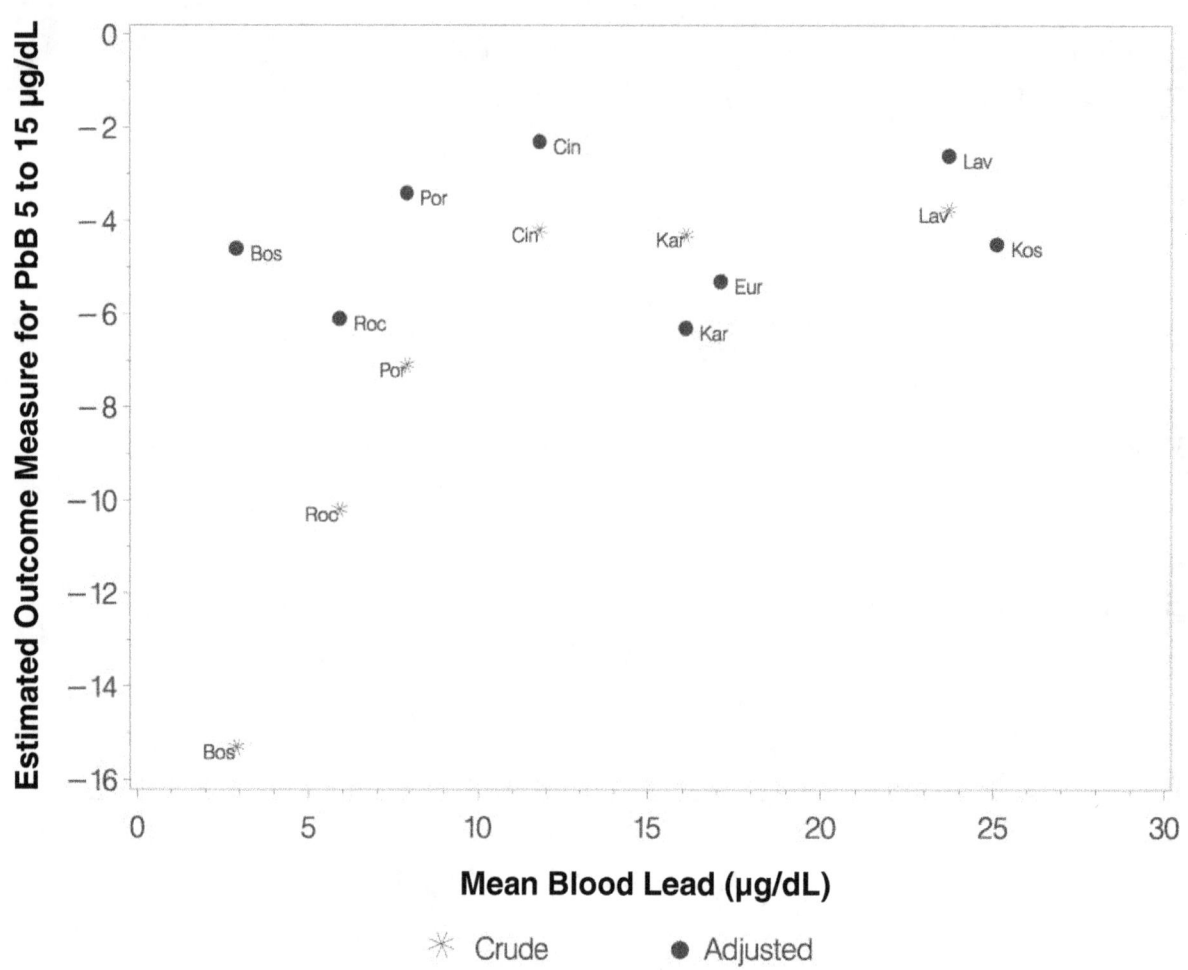

Figure 4. Age trend in blood lead levels (BLLs).

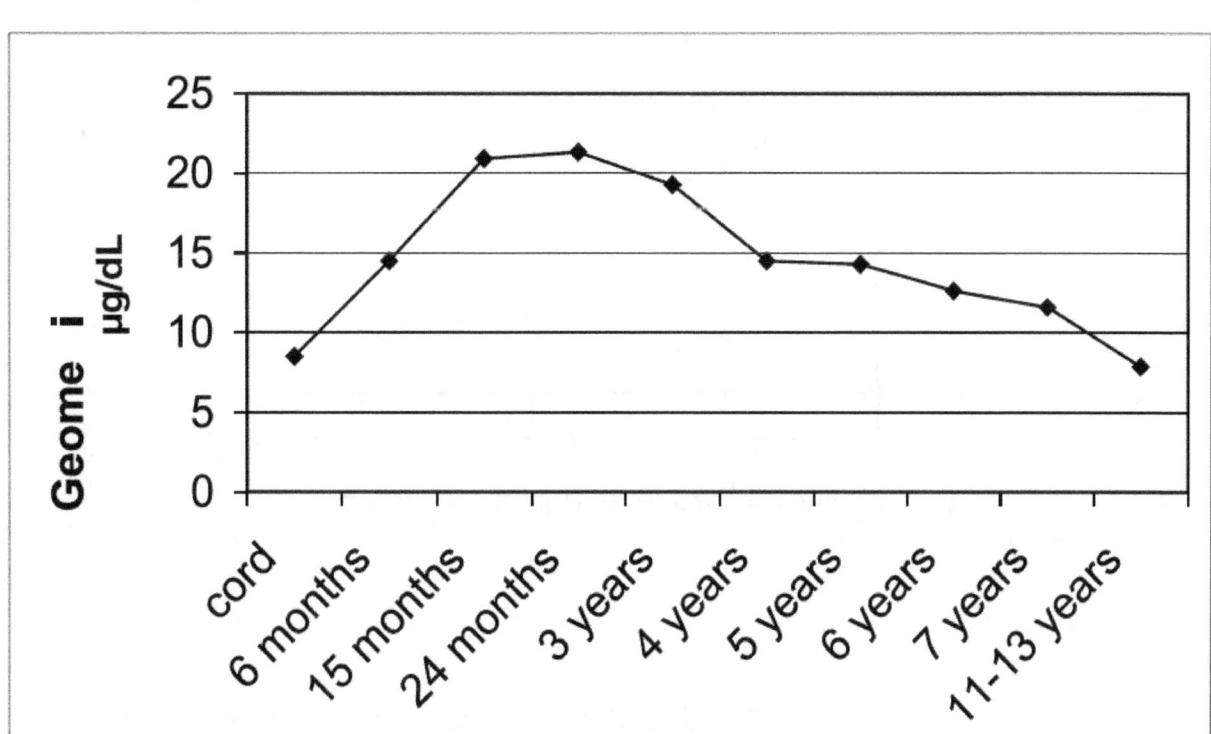

Source: Tong et al. 1996.

Figure 5. Hypothetical observed association between blood lead and IQ.

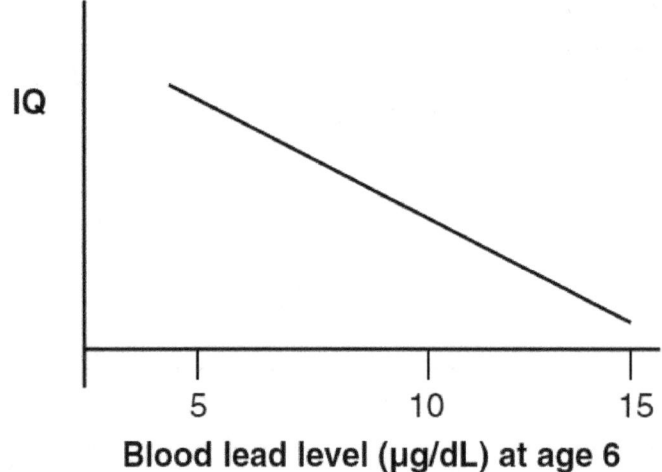

Figure 6. Hypothetical "true" association between blood lead and IQ.

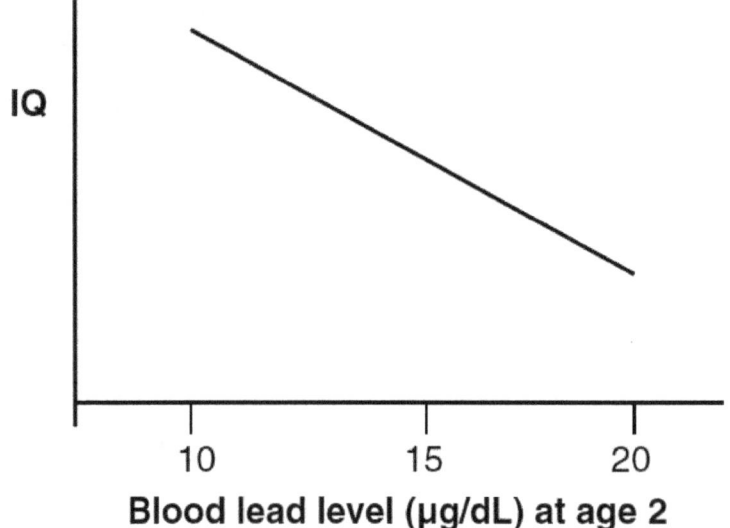

Figure 7. Hypothetical slopes of the relationship between blood lead and IQ associated with residual confounding (see Overall Conclusions).

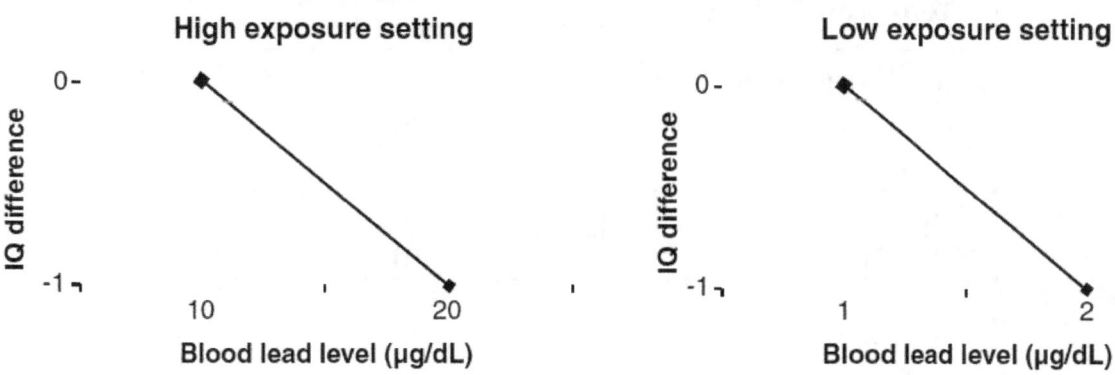

Table 1. Lowest blood lead level (BLL) considered elevated by CDC and the US Public Health Service

Year and Reference	BLL (µg/dL)
1971 (Surgeon General)	40
1975 (CDC)	30
1978 (CDC)	30
1985 (CDC)	25
1991 (CDC)	10

Table 2. Summary of studies estimating association of postnatal PbB with cognitive function

Study Population* (ref., type, n)	PbB Age	Outcome Age #	Mean PbB (ug/dL)	Estimated Delta for PbB 5 -> 15** Crude	Adj	SES	Smoking	Fetal Growth	Family Environment	HOME	Race	Parental Intelligence	Iron Status	Other
(<= 2 years)	**(>= 4 years)**													
Kosovo (37 L 332)	24 months	# 4 years	21 8 (GM)	Not stated	-5 (log10)^	Maternal Education		Birth Weight	Family Structure Maternal Age	Unspecified	Child	Maternal		Child's Sex
Port Pirie (23 L 537)	24 months	# 4 years	21 3 (GM)	-12 3 (log10)^	-5 3 (log10)^	Maternal Education Paternal Education Paternal Occupation		Birth Weight Gestation	Marital Status Maternal Age	Unspecified		Maternal		Maternal Medication/Drug Use Postnatal Factors Birth Order Birth Type Birth Problems Child's Sex Residence in Regions Child's Medical History Mother's Work Site
Port Pirie (23 L 537)	15 months	# 4 years	20 9 (GM)	-6 8 (log10)^	-1 7 (log10)	Maternal Education Paternal Education Paternal Occupation		Birth Weight Gestation	Marital Status Maternal Age	Unspecified		Maternal		Maternal Medication/Drug Use Postnatal Factors Birth Order Birth Type Birth Problems Child's Sex Residence in Regions Child's Medical History Mother's Work Site
Port Pirie (35 L 367)	15 months	11-13 years	20 9 (GM)	-6 8 (ln)^	-2 (ln)	Daniel's Scale of Prestige of Occupations in Australia Maternal Education	Parental Smoking Habits	Birth Weight	Family Structure Family Functioning Marital Status Maternal Age Life Events	Unspecified		Maternal		Breast Feeding Feeding Method Birth Order Child's Sex Child's Age School Grade School Absence

Covariates in Model

* L=Longitudinal cohort, X=Cross-sectional.
** (ln)/(log10) = Original coefficient reported in log scale.
# = McCarthy GCI, all unmarked are full-scale IQ measures.
^ statistically significant (p < 0.05)

Covariates in Model

Study Population* (ref., type, n)	PbB Age	Outcome Age #	Mean PbB (ug/dL)	Estimated Delta for PbB 5 -> 15** Crude	Adj	SES	Smoking	Fetal Growth	Family Environment	HOME	Race	Parental Intelligence	Iron Status	Other
Port Pirie (4 L 494)	Lifetime average - 2 years	7 years	16 6-20 5 (means of 2nd & 3rd quartiles) (GM)	Not stated	-5 1 (ln)^	Daniel's Scale of Prestige of Occupations in Australia Maternal Education Paternal Education	Parental Smoking	Birth Weight	Family Structure Maternal Age	Unspecified		Maternal		Breast Feeding Feeding Method Birth Order Child's Sex
Kosovo (37 L 332)	18 months	# 4 years	20 0 (GM)	Not stated	-2 3 (log10)	Maternal Education		Birth Weight	Family Structure Maternal Age	Unspecified	Child	Maternal		Child's Sex
Port Pirie (4 L 494)	Lifetime average - 15 months	7 years	14 3-18 0 (means of 2nd & 3rd quartiles) (GM)	Not stated	-4 4 (ln)^	Daniel's Scale of Prestige of Occupations in Australia Maternal Education Paternal Education	Parental Smoking	Birth Weight	Family Structure Maternal Age	Unspecified		Maternal		Breast Feeding Feeding Method Birth Order Child's Sex
Kosovo (37 L 332)	12 months	# 4 years	17 2 (GM)	Not stated	-3 6 (log10)^	Maternal Education		Birth Weight	Family Structure Maternal Age	Unspecified	Child	Maternal		Child's Sex
Cincinnati (13 L 253)	Mean 15-24 months	6 5 years	17 1	-1 3	-0 2		Cigarette Consumption during Pregnancy	Birth Weight Birth Length		Unspecified		Maternal		Child's Sex
Cleveland (16 L 149)	2 years	4 years 10 months	16 70	r=.38^	Not stated	Maternal Education	Cigarettes per Day	Birth Weight Gestation	Authoritarian Family Ideology	Total (mean of 1 2 3 and 4 years 10 months)	Child	Maternal		Child Stress Maternal Medication/Drug Use Maternal Alcohol Consumption Birth Order Child's Sex History Alcohol Abuse

* L=Longitudinal cohort, X=Cross-sectional.

** (ln)/(log10) = Original coefficient reported in log scale.

= McCarthy GCI, all unmarked are full-scale IQ measures.

^ statistically significant (p < 0.05)

Covariates in Model

Study Population* (ref, type, n)	PbB Age	Outcome Age #	Mean PbB (ug/dL)	Estimated Delta for PbB 5 -> 15**		SES	Smoking	Fetal Growth	Family Environment	HOME	Race	Parental Intelligence	Iron Status	Other
				Crude	Adj									
Sydney (12 L 318)	Mean 18 24 months	# 48 months	15 8 (GM)	Not stated	Not stated	Daniel's Scale of Prestige of Occupations in Australia Maternal Education Paternal Education		Gestation		Total at 48 months		Maternal		
Sydney (12 L 318)	Mean 6 12 months	# 48 months	15 2 (GM)	Not stated	Not stated	Daniel's Scale of Prestige of Occupations in Australia Maternal Education Paternal Education		Gestation		Total at 48 months		Maternal		
Kosovo (37 L 332)	6 months	# 4 years	15 0 (GM)	Not stated	-2 (log10)	Maternal Education		Birth Weight	Family Structure Maternal Age	Unspecified	Child	Maternal		Child's Sex
Port Pirie (23 L 537)	6 months	# 4 years	14 5 (GM)	-7 2 (log10)^	-4 1 (log10)	Maternal Education Paternal Education Paternal Occupation		Birth Weight Gestation	Marital Status Maternal Age	unspecified		Maternal		Maternal Medication/Drug Use Postnatal Factors Birth Order Birth Type Birth Problems Child's Sex Residence in Regions Child's Medical History Mother's Work Site
Costa Rica (41 L 184)	12-23 months	5 years	11 0	r= 06	Not stated									
Cincinnati (13 L 253)	Mean 3-12 months	6 5 years	10 6	-2 2	0 1		Cigarette Consumption during Pregnancy	Birth Weight Birth Length		Unspecified		Maternal		Child's Sex
Mexico City (31 L 112)	Mean 6-18 months	# 36-60 months	10 1 (GM)	Not stated	Mean square = 87 81 (neg) (ln)	Family Socioeconomic Level Maternal Education		Birth Weight				Maternal		Postnatal Factors Birth Order Child's Sex

* L=Longitudinal cohort, X=Cross-sectional.

** (ln)/(log10) = Original coefficient reported in log scale.

= McCarthy GCI, all unmarked are full-scale IQ measures.

^ statistically significant (p < 0.05)

Study Population* (ref., type, n)	PbB Age	Outcome Age #	Mean PbB (ug/dL)	Estimated Delta for PbB 5 -> 15**		SES	Smoking	Fetal Growth	Family Environment	HOME	Race	Parental Intelligence	Iron Status	Other
				Crude	Adj									
Cleveland (16 L 122)	6 months	4 years 10 months	9 99	r=-06	Not stated	Maternal Education	Cigarettes per Day	Birth Weight Gestation	Authoritarian Family Ideology	Total (mean of 1 2 3 and 4 years 10 months)	Child	Maternal		Child Stress Maternal Medication/Drug Use Maternal Alcohol Consumption Birth Order Child's Sex History Alcohol Abuse
Boston (6 L 170)	18 months	# 57 months	8 0	-3 3 (ln)^	-1 8 (ln)	Hollingshead Index of Social Class		Birth Weight	Family Structure Marital Status Residence Changes Day Care	Total	Child	Maternal		Birth Order Child's Sex Medication Used by Child Preschool Attendance
Boston (6 L 170)	12 months	# 57 months	7 8	-2 4 (ln)	-1 6 (ln)	Hollingshead Index of Social Class		Birth Weight	Family Structure Marital Status Residence Changes Day Care	Total	Child	Maternal		Birth Order Child's Sex Medication Used by Child Preschool Attendance
Boston (7 L 116)	18 months	10 years	7 8	-2 8	-1 2	Hollingshead Four-Factor Index of Social Class			Family Stress Marital Status Residence Changes Maternal Age	Scales V & V at 120 months Total at 57 months	Child	Maternal		Child Stress Birth Order Child's Sex
Boston (7 L 116)	12 months	10 years	7 7	-2	0	Hollingshead Four-Factor Index of Social Class			Family Balance Family Stress Marital Status	Scales V & V at 120 months Total at 57 months	Child	Maternal		Child Stress Parents' Sense Competence Birth Order Child's Sex
Boston (6 L 170)	24 months	# 57 months	7 0	-3 4 (ln)^	-3 2 (ln)^	Hollingshead Index of Social Class		Birth Weight	Family Structure Marital Status Residence Changes Day Care	Total	Child	Maternal		Birth Order Child's Sex Medication Used by Child Preschool Attendance

Covariates in Model

* L=Longitudinal cohort, X=Cross-sectional.
** (ln)/(log10) = Original coefficient reported in log scale.
# = McCarthy GCI, all unmarked are full-scale IQ measures.
^ statistically significant (p < 0.05)

Covariates in Model

Study Population* (ref., type, n)	PbB Age	Outcome Age #	Mean PbB (ug/dL)	Estimated Delta for PbB 5 -> 15**		SES	Smoking	Fetal Growth	Family Environment	HOME	Race	Parental Intelligence	Iron Status	Other
				Crude	Adj									
Rochester (11 L 172) [all]	Average in infancy - 6-24 months	5 years	7 0	-9 7^	-5 3^	Yearly Household Income Maternal Education	Tobacco Use during Pregnancy (user/nonuser)	Birth Weight		Total	Mother	Maternal	Serum Transferrin Saturation	Child's Sex
Boston (6 L 170)	6 months	# 57 months	6 8	0 3 (ln)	0 3 (ln)	Hollingshead Index of Social Class		Birth Weight	Family Structure Marital Status Residence Changes Day Care	Total	Child	Maternal		Birth Order Child's Sex Medication Used by Child Preschool Attendance
Boston (7 L 116)	6 months	10 years	6 7	-2	-1 3	Hollingshead Four-Factor Index of Social Class			Marital Status	Scales V & V at 120 months Total at 57 months	Child	Maternal		Child Stress Birth Order Child's Sex
Boston (7 L 116)	24 months	10 years	6 5	-7 1^	-5 8^	Hollingshead Four-Factor Index of Social Class			Marital Status Residence Changes Maternal Age	Scales V & V at 120 months Total at 57 months	Child	Maternal		Child Stress Birth Order Child's Sex
Cincinnati (13 L 253)	10 Days	6 5 years	5	-1	-0 3		Cigarette Consumption during Pregnancy	Birth Weight Birth Length		Unspecified		Maternal		Child's Sex
Boston (34 L 148)	24 months	10 years	< 8	Not stated	-5 8^	Hollingshead Four-Factor Index of Social Class			Marital Status Residence Changes Maternal Age	Scales V & V at 120 months Total at 57 months	Child	Maternal		Child Stress Birth Order Child's Sex
Rochester (11 L 105) [<10 group]	Average in infancy - 6-24 months	5 years	Not stated	-15 8^	-9 2	Yearly Household Income Maternal Education	Tobacco Use during Pregnancy (user/nonuser)	Birth Weight		Total	Mother	Maternal	Serum Transferrin Saturation	Child's Sex
(>2 - <4 years) (>= 4 years)														
Kosovo (37 L 332)	36 months	# 4 years	24 1 (GM)	Not stated	-4 5 (log10)^	Maternal Education		Birth Weight	Family Structure Maternal Age	Unspecified	Child	Maternal		Child's Sex

* L=Longitudinal cohort, X=Cross-sectional.
** (ln)/(log10) = Original coefficient reported in log scale.
# = McCarthy GCI, all unmarked are full-scale IQ measures.
^ statistically significant (p < 0.05)

45

| | | | | Estimated Delta for PbB 5 -> 15** | | Covariates in Model | | | | | | | | |
Study Population* (ref., type, n)	PbB Age	Outcome Age #	Mean PbB (ug/dL)	Crude	Adj	SES	Smoking	Fetal Growth	Family Environment	HOME	Race	Parental Intelligence	Iron Status	Other
Kosovo (37 L 332)	42 months	# 4 years	23.2 (GM)	Not stated	-5 (log10)^	Maternal Education		Birth Weight	Family Structure Maternal Age	Unspecified	Child	Maternal		Child's Sex
Kosovo (37 L 332)	30 months	# 4 years	22.1 (GM)	Not stated	-4.6 (log10)^	Maternal Education		Birth Weight	Family Structure Maternal Age	Unspecified	Child	Maternal		Child's Sex
Port Pirie (4 L 494)	Lifetime average - 3 years	7 years	17.4-21.7 (means of 2nd & 3rd quartiles) (GM)	Not stated	-5.3 (ln)^	Daniel's Scale of Prestige of Occupations in Australia Maternal Education Paternal Education	Parental Smoking	Birth Weight	Family Structure Maternal Age	Unspecified		Maternal		Breast Feeding Feeding Method Birth Order Child's Sex
Port Pirie (23 L 537)	36 months	# 4 years	19.5 (GM)	-12 (log10)^	-6.3 (log10)^	Maternal Education Paternal Education Paternal Occupation		Birth Weight Gestation	Marital Status Maternal Age	Unspecified		Maternal		Maternal Medication/Drug Use Postnatal Factors Birth Order Birth Type Birth Problems Child's Sex Residence in Regions Child's Medical History Mother's Work Site
Port Pirie (35 L 372)	3 years	11-13 years	19.3 (GM)	-10.8 (ln)^	-4.2 (ln)	Daniel's Scale of Prestige of Occupations in Australia Maternal Education	Parental Smoking Habits	Birth Weight	Family Structure Family Functioning Marital Status Maternal Age Life Events	Unspecified		Maternal		Breast Feeding Feeding Method Birth Order Child's Sex Child's Age School Grade School Absence
Cleveland (16 L 155)	3 years	4 years 10 months	16.70	r=-.31^	Not stated	Maternal Education	Cigarettes per Day	Birth Weight Gestation	Authoritarian Family Ideology	Total (mean of 1 2 3 and 4 years 10 months)	Child	Maternal		Child Stress Maternal Medication/Drug Use Maternal Alcohol Consumption Birth Order Child's Sex History Alcohol Abuse

* L=Longitudinal cohort, X=Cross-sectional.
** (ln)/(log10) = Original coefficient reported in log scale.
# = McCarthy GCI, all unmarked are full-scale IQ measures.
^ statistically significant (p < 0.05)

Covariates in Model

Study Population* (ref., type, n)	PbB Age	Outcome Age #	Mean PbB (ug/dL)	Estimated Delta for PbB 5 -> 15**		SES	Smoking	Fetal Growth	Family Environment	HOME	Race	Parental Intelligence	Iron Status	Other
				Crude	Adj									
Cincinnati (13 L 253)	Mean 27-36 months	6 5 years	16 3	-2 6^	-1 3		Cigarette Consumption during Pregnancy	Birth Weight Birth Length		Unspecified		Maternal		Child's Sex
Sydney (12 L 318)	Mean 30 36 months	# 48 months	12 4 (GM)	Not stated	Not stated	Daniel's Scale of Prestige of Occupations in Australia Maternal Education Paternal Education		Gestation		Total at 48 months		Maternal		
Cleveland (16 L 212)	Mean 0 5-3 years	4 years 10 months	9 99 at 6 months & 16 70 at both 2 years & 3 years	r=- 25	Not stated	Maternal Education	Cigarettes per Day	Birth Weight Gestation	Authoritarian Family Ideology	Total (mean of 1 2 3 and 4 years 10 months)	Child	Maternal		Child Stress Maternal Medication/Drug Use Maternal Alcohol Consumption Birth Order Child's Sex History Alcohol Abuse
Mexico City (31 L 112)	Mean 24-36 months	# 36-60 months	9 7 (GM)	Not stated	Mean square = 101 62 (neg) (ln)	Family Socioeconomic Level Maternal Education		Birth Weight				Maternal		Postnatal Factors Birth Order Child's Sex
Port Pirie (35 L 326)	Lifetime average - 3 years	11-13 years	Not stated	-10 4 (ln)^	-4 7 (ln)	Daniel's Scale of Prestige of Occupations in Australia Maternal Education	Parental Smoking Habits	Birth Weight	Family Structure Family Functioning Marital Status Maternal Age Life Events	Unspecified		Maternal		Breast Feeding Feeding Method Birth Order Child's Sex Child's Age School Grade School Absence
		(>= 4 years)	(>= 4 years)											
Kosovo (37 L 332)	48 months	# 4 years	25 1 (GM)	Not stated	-4 5 (log10)^	Maternal Education		Birth Weight	Family Structure Maternal Age	Unspecified	Child	Maternal		Child's Sex

* L=Longitudinal cohort, X=Cross-sectional.
** (ln)/(log10) = Original coefficient reported in log scale.
# = McCarthy GCI, all unmarked are full-scale IQ measures.
^ statistically significant (p < 0.05)

Study Population* (ref., type, n)	PbB Age	Outcome Age #	Mean PbB (ug/dL)	Estimated Delta for PbB 5 -> 15** Crude	Adj	SES	Smoking	Fetal Growth	Family Environment	HOME	Race	Parental Intelligence	Iron Status	Other
Lavrion Greece (20 X 509) [cov model b]	Primary school children - not specified years	Primary school children - not specified years	23 7	-3 76^	-2 66^	Maternal Education Paternal Education Paternal Occupation			Family Structure			Both		Birth Order History Alcohol Abuse Father's Age
Lavrion Greece (20 X 509) [cov model c]	Primary school children - not specified years	Primary school children - not specified years	23 7	-3 76^	-2 7^	Maternal Education Paternal Education Paternal Occupation		Birth Weight	Family Structure Marital Status Life Events			Both		Birth Order Child's Age Child's Medical History History Alcohol Abuse Father's Age Bilingualism Length of Child's Hospital Stay after Birth
Lavrion Greece (20 X 509) [cov model d]	Primary school children - not specified years	Primary school children - not specified years	23 7	-3 76^	-2 6^	Maternal Education Paternal Education Paternal Occupation		Birth Weight	Family Structure Marital Status Life Events			Both		Birth Order Child's Sex Child's Age Residence in Regions Child's Medical History History Alcohol Abuse Mouthing Behavior Father's Age Bilingualism Length of Child's Hospital Stay after Birth
Lavrion Greece (20 X 509) [cov model e]	Primary school children - not specified years	Primary school children - not specified years	23 7	-3 76^	-2 4^	Maternal Education Paternal Education Paternal Occupation		Birth Weight	Family Structure Marital Status Life Events			Both		Birth Order Child's Age School Grade Child's Medical History History Alcohol Abuse Father's Age Bilingualism Length of Child's Hospital Stay after Birth

Covariates in Model

* L=Longitudinal cohort, X=Cross-sectional.

** (ln)/(log10) = Original coefficient reported in log scale.

= McCarthy GCI, all unmarked are full-scale IQ measures.

^ statistically significant (p < 0.05)

Covariates in Model

Study Population* (ref., type, n)	PbB Age	Outcome Age #	Mean PbB (ug/dL)	Estimated Delta for PbB 5 -> 15** Crude	Adj	SES	Smoking	Fetal Growth	Family Environment	HOME	Race	Parental Intelligence	Iron Status	Other
Port Pirie (4 L 494)	Lifetime average - 4 years	7 years	17.6-21.5 (means of 2nd & 3rd quartiles) (GM)	Not stated	-5.1 (ln)	Daniel's Scale of Prestige of Occupations in Australia Maternal Education Paternal Education	Parental Smoking	Birth Weight	Family Structure Maternal Age	Unspecified		Maternal		Breast Feeding Feeding Method Birth Order Child's Sex
Port Pirie (4 L 494)	Lifetime average - 7 years	7 years	15.7-19.7 (means of 2nd & 3rd quartiles) (GM)	Not stated	-4.1 (ln)	Daniel's Scale of Prestige of Occupations in Australia Maternal Education Paternal Education	Parental Smoking	Birth Weight	Family Structure Maternal Age	Unspecified		Maternal		Breast Feeding Feeding Method Birth Order Child's Sex
Mexico City (26 X 139)	7-9 years	7-9 years	19.4	r=-.33 (ln)	r=-.32 (ln)	income Maternal Education Paternal Education								Child's Sex Type of Housing Nutritional Status (weight for height & height for age)
European Multicenter Study (39 M 1639)	6-11 years	6-11 years	17.1 (GM)	Not stated	-5.3	Maternal Education Paternal Occupation								Child's Sex Child's Age
Port Pirie (23 L 537)	48 months	# 4 years	16.4 (GM)	-9.6 (log10)^	-2.6 (log10)	Maternal Education Paternal Occupation		Birth Weight Gestation	Marital Status Maternal Age	Unspecified		Maternal		Maternal Medication/Drug Use Postnatal Factors Birth Order Birth Type Birth Problems Child's Sex Residence in Regions Child's Medical History Mother's Work Site
Karachi (28 X 138)	6-8 years	6-8 years	16.08	-4.3^	-6.3^								Haemoglobin	Child's Height for Age

* L=Longitudinal cohort, X=Cross-sectional.
** (ln)/(log10) = Original coefficient reported in log scale.
# = McCarthy GCI, all unmarked are full-scale IQ measures.
^ statistically significant (p < 0.05)

Study Population* (ref., type, n)	PbB Age	Outcome Age #	Mean PbB (ug/dL)	Estimated Delta for PbB 5 -> 15**		SES	Smoking	Fetal Growth	Family Environment	HOME	Race	Parental Intelligence	Iron Status	Other
				Crude	Adj									
Port Pirie (35 L 368)	5 years	11-13 years	14.3 (GM)	-9.8 (ln)^	-4.4 (ln)^	Daniel's Scale of Prestige of Occupations in Australia, Maternal Education	Parental Smoking Habits	Birth Weight	Family Structure, Family Functioning, Marital Status, Maternal Age, Life Events	Unspecified		Maternal		Breast Feeding, Feeding Method, Birth Order, Child's Sex, Child's Age, School Grade, School Absence
Port Pirie (35 L 326)	Lifetime average - 11-13 years	11-13 years	14.1 (GM)	-12.7 (ln)^	-4.7 (ln)^	Daniel's Scale of Prestige of Occupations in Australia, Maternal Education	Parental Smoking Habits	Birth Weight	Family Structure, Family Functioning, Marital Status, Maternal Age, Life Events	Unspecified		Maternal		Breast Feeding, Feeding Method, Birth Order, Child's Sex, Child's Age, School Grade, School Absence
Cincinnati (13 L 253)	Mean 39-48 months	6.5 years	14.0	-3.1^	-1.5		Cigarette Consumption during Pregnancy	Birth Weight, Birth Length		Unspecified		Maternal		Child's Sex
Cincinnati (13 L 253)	Mean 51-60 months	6.5 years	11.8	-4.2^	-2.3^		Cigarette Consumption during Pregnancy	Birth Weight, Birth Length		Unspecified		Maternal		Child's Sex
Port Pirie (35 L 360)	7 years	11-13 years	11.6 (GM)	-9.8 (ln)^	-3.7 (ln)^	Daniel's Scale of Prestige of Occupations in Australia, Maternal Education	Parental Smoking Habits	Birth Weight	Family Structure, Family Functioning, Marital Status, Maternal Age, Life Events	Unspecified		Maternal		Breast Feeding, Feeding Method, Birth Order, Child's Sex, Child's Age, School Grade, School Absence
Dunedin New Zealand (33 L 579)	11 years	11 years	11.1	r=-0.05 (ln)	Not stated									
Sassuolo Italy (8 X 211)	7-8 years	7-8 years	10.99 (GM)	r = -0.064 (log10)	Not stated									

* L=Longitudinal cohort, X=Cross-sectional.

** (ln)/(log10) = Original coefficient reported in log scale.

\# = McCarthy GCI, all unmarked are full-scale IQ measures.

^ statistically significant (p < 0.05)

Covariates in Model

Study Population* (ref., type, n)	PbB Age	Outcome Age #	Mean PbB (ug/dL)	Estimated Delta for PbB 5 -> 15**		SES	Smoking	Fetal Growth	Family Environment	HOME	Race	Parental Intelligence	Iron Status	Other
				Crude	Adj									
Sydney (12 L 318)	Mean 42 48 months	# 48 months	10 4 (GM)	Not stated	Not stated	Daniel's Scale of Prestige of Occupations in Australia Maternal Education Paternal Education		Gestation		Total at 48 months		Maternal		
San Luis Potosi Mexico (10 X 39) [reference group]	6-9 years	6-9 years	9 73 (GM)	r= 06 (ln)	r= 02 (ln)	Bronfman ndex of Socioeconomic Status Maternal Education Paternal Education								Child's Sex Child's Age
San Luis Potosi Mexico (10 X 41) [exposed group]	6-9 years	6-9 years	8 98 (GM)	r=- 14 (ln)	r=- 12 (ln)	Bronfman ndex of Socioeconomic Status Maternal Education Paternal Education								Child's Sex Child's Age
Mexico City (31 L 112)	Mean 42-54 months	# 42-54 months	8 4 (GM)	Not stated	Mean square = 6 23 (neg) (ln)	Family Socioeconomic Level Maternal Education		Birth Weight				Maternal		Postnatal Factors Birth Order Child's Sex
Port Pirie (35 L 326)	11-13 years	11-13 years	7 9 (GM)	-7 1 (ln)^	-3 4 (ln)^	Daniel's Scale of Prestige of Occupations in Australia Maternal Education	Parental Smoking Habits	Birth Weight	Family Structure Family Functioning Marital Status Maternal Age Life Events	Unspecified		Maternal		Breast Feeding Feeding Method Birth Order Child's Sex Child's Age School Grade School Absence
Rochester (11 L 172) [all]	Lifetime average - 5 years	5 years	7 4	-10^	-5 7^	Yearly Household ncome Maternal Education	Tobacco Use during Pregnancy (user/nonuser)	Birth Weight		Total	Mother	Maternal	Serum Transferrin Saturation	Child's Sex

* L=Longitudinal cohort, X=Cross-sectional.
** (ln)/(log10) = Original coefficient reported in log scale.
# = McCarthy GCI, all unmarked are full-scale IQ measures.
^ statistically significant (p < 0.05)

Covariates in Model

Study Population* (ref., type, n)	PbB Age	Outcome Age #	Mean PbB (ug/dL)	Estimated Delta for PbB 5 -> 15** Crude	Adj	SES	Smoking	Fetal Growth	Family Environment	HOME	Race	Parental Intelligence	Iron Status	Other
Boston (6 L 170)	57 months	# 57 months	6 4	-4 7 (ln)^	-2 5 (ln)	Hollingshead Index of Social Class		Birth Weight	Family Structure Marital Status Residence Changes Day Care	Total	Child	Maternal		Birth Order Child's Sex Medication Used by Child Preschool Attendance
Boston (7 L 116)	57 months	10 years	6 3	-9^	-2 6	Hollingshead Four-Factor Index of Social Class		Birth Weight	Family Stress Marital Status Maternal Age	Scales V & V at 120 months Total at 57 months	Child	Maternal		Child Stress Birth Order Child's Sex
Rochester (11 L 171) [all]	Concurrent - 5 years	5 years	5 9	-10 2^	-6 1^	Yearly Household Income Maternal Education	Tobacco Use during Pregnancy (user/nonuser)	Birth Weight		Total	Mother	Maternal	Serum Transferrin Saturation	Child's Sex
Boston (7 L 116)	10 years	10 years	2 9	-15 3^	-4 6	Hollingshead Four-Factor Index of Social Class		Birth Weight	Family Stress Marital Status Day Care Maternal Age	Scales V & V at 120 months Total at 57 months	Child	Maternal		Child Stress Birth Order Child's Sex
Kosovo (38 L 258)	Mean AUC7 years	7 years	PbB at age 7 years = 21 2 cumulative PbB through age 7 years = 121	-1 4 (log10)	-4 1 (log10)^	Maternal Education		Birth Weight	Family Structure Maternal Age	Unspecified	Child	Maternal		Child's Sex
Rochester (11 L 101) [<10 group]	Concurrent - 5 years	5 years	Not stated	-25 6^	-17 9^	Yearly Household Income Maternal Education	Tobacco Use during Pregnancy (user/nonuser)	Birth Weight		Total	Mother	Maternal	Serum Transferrin Saturation	Child's Sex
Rochester (11 L 101) [<10 group]	Lifetime average - 5 years	5 years	Not stated	-25 4^	-15 2^	Yearly Household Income Maternal Education	Tobacco Use during Pregnancy (user/nonuser)	Birth Weight		Total	Mother	Maternal	Serum Transferrin Saturation	Child's Sex

* L=Longitudinal cohort, X=Cross-sectional.

** (ln)/(log10) = Original coefficient reported in log scale.

= McCarthy GCI, all unmarked are full-scale IQ measures.

^ statistically significant (p < 0.05)

Covariates in Model

Study Population* (ref., type, n)	PbB Age	Outcome Age #	Mean PbB (ug/dL)	Estimated Delta for PbB 5 -> 15**		SES	Smoking	Fetal Growth	Family Environment	HOME	Race	Parental Intelligence	Iron Status	Other
				Crude	Adj									
Sydney (12 L 318)	Lifetime average - 48 months	# 48 months	Not stated	Not stated	Not stated	Daniel's Scale of Prestige of Occupations in Australia Maternal Education Paternal Education		Gestation		Total at 48 months		Maternal		
Port Pirie (35 L 326)	Lifetime average - 5 years	11-13 years	Not stated	-11 1 (ln)^	-5 6 (ln)^	Daniel's Scale of Prestige of Occupations in Australia Maternal Education	Parental Smoking Habits	Birth Weight	Family Structure Family Functioning Marital Status Maternal Age Life Events	Unspecified		Maternal		Breast Feeding Feeding Method Birth Order Child's Sex Child's Age School Grade School Absence
Port Pirie (35 L 326)	Lifetime average - 7 years	11-13 years	Not stated	-11 (ln)^	-5 1 (ln)	Daniel's Scale of Prestige of Occupations in Australia Maternal Education	Parental Smoking Habits	Birth Weight	Family Structure Family Functioning Marital Status Maternal Age Life Events	Unspecified		Maternal		Breast Feeding Feeding Method Birth Order Child's Sex Child's Age School Grade School Absence
Cincinnati (13 L 253)	Mean 66-72 months	6 5 years	Not stated	-5 8^	-3 3^		Cigarette Consumption during Pregnancy	Birth Weight Birth Length		Unspecified		Maternal		Child's Sex
Cincinnati (13 L 253)	Lifetime average - 72 months	6 5 years	Not stated	-3 1^	1 3		Cigarette Consumption during Pregnancy	Birth Weight Birth Length		Unspecified		Maternal		Child's Sex
(Other)	(Other)													
Cleveland (15 L 167)	3 years	3 years	16 95	r=- 27^	Not stated	Maternal Education			Authoritarian Family Ideology	Total	Child	Maternal		Birth Order Child's Sex Child's Age

* L=Longitudinal cohort, X=Cross-sectional.
** (ln)/(log10) = Original coefficient reported in log scale.
# = McCarthy GCI, all unmarked are full-scale IQ measures.
^ statistically significant (p < 0.05)

Covariates in Model

Study Population* (ref., type, n)	PbB Age	Outcome Age #	Mean PbB (ug/dL)	Estimated Delta for PbB 5 -> 15**		SES	Smoking	Fetal Growth	Family Environment	HOME	Race	Parental Intelligence	Iron Status	Other
				Crude	Adj									
Cleveland (14 L 153)	2 years	3 years	16 74	r=-31^	Not stated	Maternal Education	Maternal Cigarettes per Day	Birth Weight	Authoritarian Family Ideology	Preschool Inventory at 3 years	Child	Maternal		Maternal Medication/Drug Use Maternal Alcohol Consumption Birth Order Child's Sex Child's Age
Cleveland (15 L 153)	2 years	3 years	16 74	r=-31^	Not stated	Maternal Education			Authoritarian Family Ideology	Total	Child	Maternal		Birth Order Child's Sex Child's Age
Cleveland (14 L 165)	3 years	3 years	16 68	r=-29^	Not stated	Maternal Education	Maternal Cigarettes per Day	Birth Weight	Authoritarian Family Ideology	Preschool Inventory at 3 years	Child	Maternal		Maternal Medication/Drug Use Maternal Alcohol Consumption Birth Order Child's Sex Child's Age
Rochester (11 L 172) [all]	Peak - 5 years	5 years	11 3	-4 7^	-2 6^	Yearly Household Income Maternal Education	Tobacco Use during Pregnancy (user/nonuser)	Birth Weight		Total	Mother	Maternal	Serum Transferrin Saturation	Child's Sex
Cleveland (14 L 126)	6 months	3 years	10 05	r=-04	Not stated	Maternal Education	Maternal Cigarettes per Day	Birth Weight	Authoritarian Family Ideology	Preschool Inventory at 3 years	Child	Maternal		Maternal Medication/Drug Use Maternal Alcohol Consumption Birth Order Child's Sex Child's Age
Cleveland (15 L 126)	6 months	3 years	10 05	r=-04	Not stated	Maternal Education			Authoritarian Family Ideology	Total	Child	Maternal		Birth Order Child's Sex Child's Age
Rochester (11 L 172) [all]	Lifetime average - 3 years	3 years	7 7	-7 4^	-3 5	Yearly Household Income Maternal Education	Tobacco Use during Pregnancy (user/nonuser)	Birth Weight		Total	Mother	Maternal	Serum Transferrin Saturation	Child's Sex

* L=Longitudinal cohort, X=Cross-sectional.
** (ln)/(log10) = Original coefficient reported in log scale.
# = McCarthy GCI, all unmarked are full-scale IQ measures.
^ statistically significant (p < 0.05)

54

Covariates in Model

Study Population* (ref, type, n)	PbB Age	Outcome Age #	Mean PbB (ug/dL)	Estimated Delta for PbB 5 -> 15**		SES	Smoking	Fetal Growth	Family Environment	HOME	Race	Parental Intelligence	Iron Status	Other
				Crude	Adj									
Rochester (11 L 172) [all]	Average in infancy - 6-24 months	3 years	7.0	-7.3^	-3.2	Yearly Household Income Maternal Education	Tobacco Use during Pregnancy (user/nonuser)	Birth Weight		Total	Mother	Maternal	Serum Transferrin Saturation	Child's Sex
Rochester (11 L 172) [all]	Average in infancy - 6-24 months	3 & 5 years	7.0	-8.5^	-4.3^	Yearly Household Income Maternal Education	Tobacco Use during Pregnancy (user/nonuser)	Birth Weight		Total	Mother	Maternal	Serum Transferrin Saturation	Child's Sex
Rochester (11 L 172) [all]	Lifetime average - 3 & 5 years	3 & 5 years	Not stated	-8.7^	-4.6^	Yearly Household Income Maternal Education	Tobacco Use during Pregnancy (user/nonuser)	Birth Weight		Total	Mother	Maternal	Serum Transferrin Saturation	Child's Sex
Rochester (11 L 172) [all]	Peak - 3 years	3 years	Not stated	-4^	-1.9	Yearly Household Income Maternal Education	Tobacco Use during Pregnancy (user/nonuser)	Birth Weight		Total	Mother	Maternal	Serum Transferrin Saturation	Child's Sex
Rochester (11 L 171) [all]	Concurrent - 3 years	3 years	Not stated	-6^	-3.1^	Yearly Household Income Maternal Education	Tobacco Use during Pregnancy (user/nonuser)	Birth Weight		Total	Mother	Maternal	Serum Transferrin Saturation	Child's Sex
Rochester (11 L 171) [all]	Concurrent - 3 & 5 years	3 & 5 years	Not stated	-8.1^	-4.6^	Yearly Household Income Maternal Education	Tobacco Use during Pregnancy (user/nonuser)	Birth Weight		Total	Mother	Maternal	Serum Transferrin Saturation	Child's Sex
Rochester (11 L 172) [all]	Peak - 3 & 5 years	3 & 5 years	Not stated	-4.4^	-2.3^	Yearly Household Income Maternal Education	Tobacco Use during Pregnancy (user/nonuser)	Birth Weight		Total	Mother	Maternal	Serum Transferrin Saturation	Child's Sex

* L=Longitudinal cohort, X=Cross-sectional.
** (ln)/(log10) = Original coefficient reported in log scale.
# = McCarthy GCI, all unmarked are full-scale IQ measures.
^ = statistically significant (p < 0.05)

Covariates in Model

Study Population* (ref., type, n)	PbB Age	Outcome Age #	Mean PbB (ug/dL)	Estimated Delta for PbB 5 -> 15**		SES	Smoking	Fetal Growth	Family Environment	HOME	Race	Parental Intelligence	Iron Status	Other
				Crude	Adj									
Rochester (11 L 101) [<10 group]	Lifetime average - 3 years	3 years	Not stated	-23^	-12 2	Yearly Household Income Maternal Education	Tobacco Use during Pregnancy (user/nonuser)	Birth Weight		Total	Mother	Maternal	Serum Transferrin Saturation	Child's Sex
Rochester (11 L 101) [<10 group]	Lifetime average - 3 & 5 years	3 & 5 years	Not stated	-24 2^	-13 7^	Yearly Household Income Maternal Education	Tobacco Use during Pregnancy (user/nonuser)	Birth Weight		Total	Mother	Maternal	Serum Transferrin Saturation	Child's Sex
Rochester (11 L 101) [<10 group]	Peak - 3 years	3 years	Not stated	-20 9^	-13 6^	Yearly Household Income Maternal Education	Tobacco Use during Pregnancy (user/nonuser)	Birth Weight		Total	Mother	Maternal	Serum Transferrin Saturation	Child's Sex
Rochester (11 L 101) [<10 group]	Peak - 5 years	5 years	Not stated	-21 2^	-14 4^	Yearly Household Income Maternal Education	Tobacco Use during Pregnancy (user/nonuser)	Birth Weight		Total	Mother	Maternal	Serum Transferrin Saturation	Child's Sex
Rochester (11 L 101) [<10 group]	Peak - 3 & 5 years	3 & 5 years	Not stated	-21^	-14^	Yearly Household Income Maternal Education	Tobacco Use during Pregnancy (user/nonuser)	Birth Weight		Total	Mother	Maternal	Serum Transferrin Saturation	Child's Sex
Rochester (11 L 101) [<10 group]	Concurrent - 3 & 5 years	3 & 5 years	Not stated	-23 8^	-15 8^	Yearly Household Income Maternal Education	Tobacco Use during Pregnancy (user/nonuser)	Birth Weight		Total	Mother	Maternal	Serum Transferrin Saturation	Child's Sex
Rochester (11 L 105) [<10 group]	Average in infancy - 6-24 months	3 years	Not stated	-12 9	-5 8	Yearly Household Income Maternal Education	Tobacco Use during Pregnancy (user/nonuser)	Birth Weight		Total	Mother	Maternal	Serum Transferrin Saturation	Child's Sex

* L=Longitudinal cohort, X=Cross-sectional.

** (ln)/(log10) = Original coefficient reported in log scale.

= McCarthy GCI, all unmarked are full-scale IQ measures.

^ statistically significant (p < 0.05)

Covariates in Model

Study Population* (ref, type, n)	PbB Age	Outcome Age #	Mean PbB (ug/dL)	Estimated Delta for PbB 5 -> 15** Crude	Adj	SES	Smoking	Fetal Growth	Family Environment	HOME	Race	Parental Intelligence	Iron Status	Other
Rochester (11 L 105) [<10 group]	Average in infancy - 6-24 months	3 & 5 years	Not stated	-14.3^	-7.5	Yearly Household Income Maternal Education	Tobacco Use during Pregnancy (user/nonuser)	Birth Weight		Total	Mother	Maternal	Serum Transferrin Saturation	Child's Sex
Rochester (11 L 101) [<10 group]	Concurrent - 3 years	3 years	Not stated	-21.9^	-13.6^	Yearly Household Income Maternal Education	Tobacco Use during Pregnancy (user/nonuser)	Birth Weight		Total	Mother	Maternal	Serum Transferrin Saturation	Child's Sex

* L=Longitudinal cohort, X=Cross-sectional.

** (ln)/(log10) = Original coefficient reported in log scale.

= McCarthy GCI, all unmarked are full-scale IQ measures.

^ statistically significant (p < 0.05)

Table 3. Summary of studies estimating association of postnatal PbB with performance scale IQ

Study Population* (ref., type, n)	PbB Age (<= 2 years)	Outcome Age (>= 4 years)	Mean PbB (ug/dL)	Estimated Delta IQ for PbB 5 -> 15** Crude	Adj	SES	Smoking	Fetal Growth	Family Environment	HOME	Race	Parental IQ	Iron Status	Other
Port Pirie (35 L 367)	15 months	11-13 years	20 9 (GM)	-5 7 (ln)^	-0 7 (ln)	Daniel's Scale of Prestige of Occupations in Australia Maternal Education	Parental Smoking Habits	Birth Weight	Family Structure Family Functioning Marital Status Maternal Age Life Events	Unspecified		Maternal		Breast Feeding Feeding Method Birth Order Child's Sex Child's Age School Grade School Absence
Port Pirie (4 L 494)	Lifetime average - 2 years	7 years	16 6-20 5 (means of 2nd & 3rd quartiles) (GM)	Not stated	-2 6 (ln)	Daniel's Scale of Prestige of Occupations in Australia Maternal Education Paternal Education	Parental Smoking	Birth Weight	Family Structure Maternal Age	Unspecified		Maternal		Breast Feeding Feeding Method Birth Order Child's Sex
Port Pirie (4 L 494)	Lifetime average - 15 months	7 years	14 3-18 0 (means of 2nd & 3rd quartiles) (GM)	Not stated	-2 5 (ln)	Daniel's Scale of Prestige of Occupations in Australia Maternal Education Paternal Education	Parental Smoking	Birth Weight	Family Structure Maternal Age	Unspecified		Maternal		Breast Feeding Feeding Method Birth Order Child's Sex
Cincinnati (13 L 253)	Mean 15-24 months	6 5 years	17 1	-2^	-1		Cigarette Consumption during Pregnancy	Birth Weight Birth Length		Unspecified		Maternal		Child's Sex
Cleveland (16 L 149)	2 years	4 years 10 months	16 70	r=-.34	Not stated	Maternal Education	Cigarettes per Day	Birth Weight Gestation	Authoritarian Family Ideology	Total (mean of 1 2 3 and 4 years 10 months)	Child	Maternal		Child Stress Maternal Medication/Drug Use Maternal Alcohol Consumption Birth Order Child's Sex History Alcohol Abuse

Covariates in Model

* L=Longitudinal cohort, X=Cross-sectional.

** (ln)/(log10) = Original coefficient reported in log scale.

^ statistically significant (p < 0.05)

Covariates in Model

Study Population* (ref., type, n)	PbB Age	Outcome Age	Mean PbB (ug/dL)	Estimated Delta IQ for PbB 5 -> 15** Crude	Adj	SES	Smoking	Fetal Growth	Family Environment	HOME	Race	Parental IQ	Iron Status	Other
Costa Rica (41 L 184)	12-23 months	5 years	11.0	r=.05	Not stated									Child's Sex
Cincinnati (13 L 253)	Mean 3-12 months	6.5 years	10.6	-3.9^	-1.6		Cigarette Consumption during Pregnancy	Birth Weight Birth Length		Unspecified		Maternal		Child's Sex
Cleveland (16 L 122)	6 months	4 years 10 months	9.99	r=-.06	Not stated	Maternal Education	Cigarettes per Day	Birth Weight Gestation	Authoritarian Family Ideology	Total (mean of 1 2 3 and 4 years 10 months)	Child	Maternal		Child Stress Maternal Medication/Drug Use Maternal Alcohol Consumption Birth Order Child's Sex History Alcohol Abuse
Boston (7 L 116)	18 months	10 years	7.8	Not stated	0	Hollingshead Four-Factor Index of Social Class			Family Stress Marital Status Residence Changes Maternal Age	Scales V & V at 120 months Total at 57 months	Child	Maternal		Child Stress Birth Order Child's Sex
Boston (7 L 116)	12 months	10 years	7.7	Not stated	1.4	Hollingshead Four-Factor Index of Social Class			Family Balance Family Stress Marital Status	Scales V & V at 120 months Total at 57 months	Child	Maternal		Child Stress Parents' Sense Competence Birth Order Child's Sex
Boston (7 L 116)	6 months	10 years	6.7	Not stated	0.3	Hollingshead Four-Factor Index of Social Class			Marital Status	Scales V & V at 120 months Total at 57 months	Child	Maternal		Child Stress Birth Order Child's Sex
Boston (7 L 116)	24 months	10 years	6.5	Not stated	-3.9	Hollingshead Four-Factor Index of Social Class			Family Stress Marital Status Residence Changes Maternal Age	Scales V & V at 120 months Total at 57 months	Child	Maternal		Child Stress Birth Order Child's Sex
Cincinnati (13 L 253)	10 Days	6.5 years	5	-.4	-2.2		Cigarette Consumption during Pregnancy	Birth Weight Birth Length		Unspecified		Maternal		Child's Sex

* L=Longitudinal cohort, X=Cross-sectional.
** (ln)/(log10) = Original coefficient reported in log scale.
^ statistically significant (p < 0.05)

Covariates in Model

Study Population* (ref., type, n)	PbB Age	Outcome Age	Mean PbB (ug/dL)	Estimated Delta IQ for PbB 5 -> 15** Crude	Adj	SES	Smoking	Fetal Growth	Family Environment	HOME	Race	Parental IQ	Iron Status	Other
Boston (34 L 148)	24 months	10 years	< 8	Not stated	-3 9	Hollingshead Four-Factor Index of Social Class			Marital Status Residence Changes Maternal Age	Scales V & V at 120 months Total at 57 mo	Child	Maternal		Child Stress Birth Order Child's Sex
(>2 - <4 years) (>= 4 years)														
Port Pirie (4 L 494)	Lifetime average - 3 years	7 years	17 4-21 7 (means of 2nd & 3rd quartiles) (GM)	Not stated	-3 1 (ln)	Daniel's Scale of Prestige of Occupations in Australia Maternal Education Paternal Education	Parental Smoking	Birth Weight	Family Structure Maternal Age	Unspecified		Maternal		Breast Feeding Feeding Method Birth Order Child's Sex
Port Pirie (35 L 372)	3 years	11-13 years	19 3 (GM)	-10 3 (ln)^	-4 6 (ln)	Daniel's Scale of Prestige of Occupations in Australia Maternal Education	Parental Smoking Habits	Birth Weight	Family Structure Family Functioning Marital Status Maternal Age Life Events	Unspecified		Maternal		Breast Feeding Feeding Method Birth Order Child's Sex Child's Age School Grade School Absence
Cleveland (16 L 155)	3 years	4 years 10 months	16 70	r=- 28	Not stated	Maternal Education	Cigarettes per Day	Birth Weight Gestation	Authoritarian Family Ideology	Total (mean of 1 2 3 and 4 years 10 months)	Child	Maternal		Child Stress Maternal Medication/Drug Use Maternal Alcohol Consumption Birth Order Child's Sex History Alcohol Abuse
Cincinnati (13 L 253)	Mean 27-36 months	6 5 years	16 3	-3 4^	-2 2^		Cigarette Consumption during Pregnancy	Birth Weight Birth Length		Unspecified		Maternal		Child's Sex

* L=Longitudinal cohort, X=Cross-sectional.
** (ln)/(log10) = Original coefficient reported in log scale.
^ statistically significant (p < 0.05)

Covariates in Model

Study Population* (ref., type, n)	PbB Age	Outcome Age	Mean PbB (ug/dL)	Estimated Delta IQ for PbB 5 -> 15** Crude	Estimated Delta IQ for PbB 5 -> 15** Adj	SES	Smoking	Fetal Growth	Family Environment	HOME	Race	Parental IQ	Iron Status	Other
Cleveland (16 L 212)	Mean 0.5-3 years	4 years 10 months	9.99 at 6 months & 16.70 at both 2 years & 3 years	r=-.25	Not stated	Maternal Education	Cigarettes per Day	Birth Weight Gestation	Authoritarian Family Ideology	Total (mean of 1 2 3 and 4 years 10 months)	Child	Maternal		Child Stress Maternal Medication/Drug Use Maternal Alcohol Consumption Birth Order Child's Sex History Alcohol Abuse
Port Pirie (35 L 326)	Lifetime average - 3 years	11-13 years	Not stated	-8.6 (ln)^	-3.5 (ln)	Daniel's Scale of Prestige of Occupations in Australia Maternal Education	Parental Smoking Habits	Birth Weight	Family Structure Family Functioning Marital Status Maternal Age Life Events	Unspecified		Maternal		Breast Feeding Feeding Method Birth Order Child's Sex Child's Age School Grade School Absence
(>= 4 years)	**(>= 4 years)**													
Lavrion Greece (20 X 509)	Primary school children - not specified years	Primary school children - not specified years	23.7	Not stated	-2.3^	Maternal Education Paternal Education Paternal Occupation		Birth Weight	Family Structure Marital Status Life Events			Both		Birth Order Child's Age Child's Medical History Alcohol Abuse Father's Age Bilingualism Length of Child's Hospital Stay after Birth
Port Pirie (4 L 494)	Lifetime average - 4 years	7 years	17.6-21.5 (means of 2nd & 3rd quartiles) (GM)	Not stated	-3.6 (ln)	Daniel's Scale of Prestige of Occupations in Australia Maternal Education Paternal Education	Parental Smoking	Birth Weight	Family Structure Maternal Age	Unspecified		Maternal		Breast Feeding Feeding Method Birth Order Child's Sex

* L=Longitudinal cohort, X=Cross-sectional.

** (ln)/(log10) = Original coefficient reported in log scale.

^ statistically significant (p < 0.05)

Covariates in Model

Study Population* (ref., type, n)	PbB Age	Outcome Age	Mean PbB (ug/dL)	Estimated Delta IQ for PbB 5 -> 15** Crude	Adj	SES	Smoking	Fetal Growth	Family Environment	HOME	Race	Parental IQ	Iron Status	Other
Port Pirie (4 L 494)	Lifetime average - 7 years	7 years	15 7-19 7 (means of 2nd & 3rd quartiles) (GM)	Not stated	-2 5 (ln)	Daniel's Scale of Prestige of Occupations in Australia Maternal Education Paternal Education	Parental Smoking	Birth Weight	Family Structure Maternal Age	Unspecified		Maternal		Breast Feeding Feeding Method Birth Order Child's Sex
Mexico City (26 X 139)	7-9 years	7-9 years	19 4	r=- 24 (ln)	r=- 28 (ln)	ncome Maternal Education Paternal Education								Child's Sex Type of Housing Nutritional Status (weight for height & height for age)
Port Pirie (35 L 368)	5 years	11-13 years	14 3 (GM)	-7 9 (ln)^	-4 1 (ln)	Daniel's Scale of Prestige of Occupations in Australia Maternal Education	Parental Smoking Habits	Birth Weight	Family Structure Family Functioning Marital Status Maternal Age Life Events	Unspecified		Maternal		Breast Feeding Feeding Method Birth Order Child's Sex Child's Age School Grade School Absence
Port Pirie (35 L 326)	Lifetime average - 11-13 years	11-13 years	14 1 (GM)	-11 9 (ln)^	-5 2 (ln)	Daniel's Scale of Prestige of Occupations in Australia Maternal Education	Parental Smoking Habits	Birth Weight	Family Structure Family Functioning Marital Status Maternal Age Life Events	Unspecified		Maternal		Breast Feeding Feeding Method Birth Order Child's Sex Child's Age School Grade School Absence
Cincinnati (13 L 253)	Mean 39-48 months	6 5 years	14 0	-4 3^	-2 7^		Cigarette Consumption during Pregnancy	Birth Weight Birth Length		Unspecified		Maternal		Child's Sex
Cincinnati (13 L 253)	Mean 51-60 months	6 5 years	11 8	-5 5^	-3 8^		Cigarette Consumption during Pregnancy	Birth Weight Birth Length		Unspecified		Maternal		Child's Sex

* L=Longitudinal cohort, X=Cross-sectional.
** (ln)/(log10) = Original coefficient reported in log scale.
^ statistically significant (p < 0.05)

Covariates in Model

Study Population* (ref., type, n)	PbB Age	Outcome Age	Mean PbB (ug/dL)	Estimated Delta IQ for PbB 5 -> 15** Crude	Adj	SES	Smoking	Fetal Growth	Family Environment	HOME	Race	Parental IQ	Iron Status	Other
Port Pirie (35 L 360)	7 years	11-13 years	11 6 (GM)	-9 4 (ln)^	-4 2 (ln)	Daniel's Scale of Prestige of Occupations in Australia Maternal Education	Parental Smoking Habits	Birth Weight	Family Structure Family Functioning Marital Status Maternal Age Life Events	Unspecified		Maternal		Breast Feeding Feeding Method Birth Order Child's Sex Child's Age School Grade School Absence
Dunedin New Zealand (33 L 579)	11 years	11 years	11 1	r=-0 03 (ln)	Not stated									
Sassuolo Italy (8 X 211)	7-8 years	7-8 years	10 99 (GM)	r = -0 100 (log10)	Not stated									
San Luis Potosi Mexico (10 X 39) [reference group]	6-9 years	6-9 years	9 73 (GM)	r= 04 (ln)	r=-.10 (ln)	Bronffman ndex of Socioeconomic Status Maternal Education Paternal Education								Child's Sex Child's Age
San Luis Potosi Mexico (10 X 41) [exposed group]	6-9 years	6-9 years	8 98 (GM)	r=-.08 (ln)	r=.005 (ln)	Bronffman ndex of Socioeconomic Status Maternal Education Paternal Education								Child's Sex Child's Age
Port Pirie (35 L 326)	11-13 years	11-13 years	7 9 (GM)	-6 8 (ln)^	-2 2 (ln)	Daniel's Scale of Prestige of Occupations in Australia Maternal Education	Parental Smoking Habits	Birth Weight	Family Structure Family Functioning Marital Status Maternal Age Life Events	Unspecified		Maternal		Breast Feeding Feeding Method Birth Order Child's Sex Child's Age School Grade School Absence
Boston (7 L 116)	57 months	10 years	6 3	Not stated	-4 4	Hollingshead Four-Factor ndex of Social Class		Birth Weight	Family Stress Marital Status Maternal Age	Scales V & V at 120 months Total at 57 months	Child	Maternal		Child Stress Birth Order Child's Sex

* L=Longitudinal cohort, X=Cross-sectional.
** (ln)/(log10) = Original coefficient reported in log scale.
^ statistically significant (p < 0.05)

Covariates in Model

Study Population* (ref., type, n)	PbB Age	Outcome Age	Mean PbB (ug/dL)	Estimated Delta IQ for PbB 5 -> 15**		SES	Smoking	Fetal Growth	Family Environment	HOME	Race	Parental IQ	Iron Status	Other
				Crude	Adj									
Boston (7 L 116)	10 years	10 years	2 9	Not stated	-1 7	Hollingshead Four-Factor Index of Social Class		Birth Weight	Family Stress Marital Status Day Care Maternal Age	Scales V & V at 120 months Total at 57 months	Child	Maternal		Child Stress Birth Order Child's Sex
Kosovo (38 L 261)	Mean AUC7 years	7 years	PbB at age 7 years = 21 2 cumulative PbB through age 7 years = 1 21	Not stated	-4 5 (log10)^	Maternal Education		Birth Weight	Family Structure Maternal Age	Unspecified	Child	Maternal		Child's Sex
Port Pirie (35 L 326)	Lifetime average - 7 years	11-13 years	Not stated	-9 6 (ln)^	-4 7 (ln)	Daniel's Scale of Prestige of Occupations in Australia Maternal Education	Parental Smoking Habits	Birth Weight	Family Structure Family Functioning Marital Status Maternal Age Life Events	Unspecified		Maternal		Breast Feeding Feeding Method Birth Order Child's Sex Child's Age School Grade School Absence
Port Pirie (35 L 326)	Lifetime average - 5 years	11-13 years	Not stated	-9 4 (ln)^	-4 8 (ln)	Daniel's Scale of Prestige of Occupations in Australia Maternal Education	Parental Smoking Habits	Birth Weight	Family Structure Family Functioning Marital Status Maternal Age Life Events	Unspecified		Maternal		Breast Feeding Feeding Method Birth Order Child's Sex Child's Age School Grade School Absence
Cincinnati (13 L 253)	Mean 66-72 months	6 5 years	Not stated	-7 5^	-5 2^		Cigarette Consumption during Pregnancy	Birth Weight Birth Length		Unspecified		Maternal		Child's Sex
Cincinnati (13 L 253)	Lifetime average - 72 months	6 5 years	Not stated	-4 3^	-2 6^		Cigarette Consumption during Pregnancy	Birth Weight Birth Length		Unspecified		Maternal		Child's Sex

* L=Longitudinal cohort; X=Cross-sectional.

** (ln)/(log10) = Original coefficient reported in log scale.

^ statistically significant (p < 0.05)

Table 4. Summary of studies estimating association of postnatal PbB with verbal scale IQ

Study Population* (ref., type, n)	PbB Age	Outcome Age	Mean PbB (ug/dL)	Estimated Delta IQ for PbB 5 -> 15**		Covariates in Model								
				Crude	Adj	SES	Smoking	Fetal Growth	Family Environment	HOME	Race	Parental IQ	Iron Status	Other
	(<= 2 years)	(>= 4 years)												
Port Pirie (35 L 367)	15 months	11-13 years	20 9 (3M)	-7 (ln)^	-3 2 (ln)^	Daniel's Scale of Prestige of Occupations in Australia / Maternal Education	Parental Smoking Habits	Birth Weight	Family Structure / Family Functioning / Marital Status / Maternal Age / Life Events	Unspecified		Maternal		Breast Feeding / Feeding Method / Birth Order / Child's Sex / Child's Age / School Grade / School Absence
Port Pirie (4 L 494)	Lifetime average - 2 years	7 years	16 6-20 5 (means of 2nd & 3rd quartiles) (GM)	Not stated	-6 4 (ln)^	Daniel's Scale of Prestige of Occupations in Australia / Maternal Education / Paternal Education	Parental Smoking	Birth Weight	Family Structure / Maternal Age	Unspecified		Maternal		Breast Feeding / Feeding Method / Birth Order / Child's Sex
Port Pirie (4 L 494)	Lifetime average - 15 months	7 years	14 3-18 0 (means of 2nd & 3rd quartiles) (GM)	Not stated	-5 5 (ln)^	Daniel's Scale of Prestige of Occupations in Australia / Maternal Education / Paternal Education	Parental Smoking	Birth Weight	Family Structure / Maternal Age	Unspecified		Maternal		Breast Feeding / Feeding Method / Birth Order / Child's Sex
Cincinnati (13 L 253)	Mean 15-24 months	6 5 years	17 1	-0 3	0 2		Cigarette Consumption during Pregnancy	Birth Weight / Birth Length		Unspecified		Maternal		Child's Sex
Cleveland (16 L 149)	2 years	4 years 10 months	16 70	r=-37	Not stated	Maternal Education	Cigarettes per Day	Birth Weight / Gestation	Authoritarian Family Ideology	Total (mean of 1 2 3 and 4 years 10 months)	Child	Maternal		Child Stress / Maternal Medication/Drug Use / Maternal Alcohol Consumption / Birth Order / Child's Sex / History Alcohol Abuse

* L=Longitudinal cohort, X=Cross-sectional.
** (ln)/(log10) = Original coefficient reported in log scale.
^ statistically significant (p < 0.05)

Covariates in Model

Study Population* (ref, type, n)	PbB Age	Outcome Age	Mean PbB (ug/dL)	Estimated Delta IQ for PbB 5 -> 15**		SES	Smoking	Fetal Growth	Family Environment	HOME	Race	Parental IQ	Iron Status	Other
				Crude	Adj									
Costa Rica (41 L 184)	12-23 months	5 years	11.0	r=.06	Not stated									Child's Sex
Cincinnati (13 L 253)	Mean 3-12 months	6.5 years	10.6	0	1.2		Cigarette Consumption during Pregnancy	Birth Weight Birth Length		Unspecified		Maternal		Child's Sex
Cleveland (16 L 122)	6 months	4 years 10 months	9.99	r=-.05	Not stated	Maternal Education	Cigarettes per Day	Birth Weight Gestation	Authoritarian Family Ideology	Total (mean of 1 2 3 and 4 years 10 months)	Child	Maternal		Child Stress Maternal Medication/Drug Use Maternal Alcohol Consumption Birth Order Child's Sex History Alcohol Abuse
Boston (7 L 116)	18 months	10 years	7.8	Not stated	-2	Hollingshead Four-Factor Index of Social Class			Family Stress Marital Status Residence Changes Maternal Age	Scales V & V at 120 months Total at 57 months	Child	Maternal		Child Stress Birth Order Child's Sex
Boston (7 L 116)	12 months	10 years	7.7	Not stated	-1.3	Hollingshead Four-Factor Index of Social Class			Family Balance Family Stress Marital Status	Scales V & V at 120 months Total at 57 months	Child	Maternal		Child Stress Parents' Sense Competence Birth Order Child's Sex
Boston (7 L 116)	6 months	10 years	6.7	Not stated	-2.4	Hollingshead Four-Factor Index of Social Class			Marital Status	Scales V & V at 120 months Total at 57 months	Child	Maternal		Child Stress Birth Order Child's Sex
Boston (7 L 116)	24 months	10 years	6.5	Not stated	-6.3^	Hollingshead Four-Factor Index of Social Class			Marital Status Residence Changes Maternal Age	Scales V & V at 120 months Total at 57 months	Child	Maternal		Child Stress Birth Order Child's Sex
Cincinnati (13 L 253)	10 Days	6.5 years	5	-.01	1.1		Cigarette Consumption during Pregnancy	Birth Weight Birth Length		Unspecified		Maternal		Child's Sex

* L=Longitudinal cohort, X=Cross-sectional.
** (ln)/(log10) = Original coefficient reported in log scale.
^ statistically significant (p < 0.05)

Covariates in Model

Study Population* (ref., type, n)	PbB Age	Outcome Age	Mean PbB (ug/dL)	Estimated Delta IQ for PbB 5 -> 15** Crude	Adj	SES	Smoking	Fetal Growth	Family Environment	HOME	Race	Parental IQ	Iron Status	Other
Boston (34 L 148)	24 months	10 years	< 8	Not stated	-6.3^	Hollingshead Four-Factor Index of Social Class			Marital Status Residence Changes Maternal Age	Scales V & V at 120 months Total at 57 months	Child	Maternal		Child Stress Birth Order Child's Sex
(>2 - <4 years) (>= 4 years)														
Port Pirie (4 L 494)	Lifetime average - 3 years	7 years	17.4-21.7 (means of 2nd & 3rd quartiles) (GM)	Not stated	-6.3 (ln)^	Daniel's Scale of Prestige of Occupations in Australia Maternal Education Paternal Education	Parental Smoking	Birth Weight	Family Structure Maternal Age	Unspecified		Maternal		Breast Feeding Feeding Method Birth Order Child's Sex
Port Pirie (35 L 372)	3 years	11-13 years	19.3 (GM)	-9.3 (ln)^	-2.9 (ln)	Daniel's Scale of Prestige of Occupations in Australia Maternal Education	Parental Smoking Habits	Birth Weight	Family Structure Family Functioning Marital Status Maternal Age Life Events	Unspecified		Maternal		Breast Feeding Feeding Method Birth Order Child's Age Child's Sex School Grade School Absence
Cleveland (16 L 155)	3 years	4 years 10 months	16.70	r=-.37	Not stated	Maternal Education	Cigarettes per Day	Birth Weight Gestation	Authoritarian Family Ideology	Total (mean of 1 2 3 and 4 years 10 months)	Child	Maternal		Child Stress Maternal Medication/Drug Use Maternal Alcohol Consumption Birth Order Child's Sex History Alcohol Abuse
Cincinnati (13 L 253)	Mean 27-36 months	6.5 years	16.3		-0.4	Cigarette Consumption during Pregnancy		Birth Weight Birth Length		Unspecified		Maternal		Child's Sex

* L=Longitudinal cohort, X=Cross-sectional.

** (ln)/(log10) = Original coefficient reported in log scale.

^ statistically significant (p < 0.05)

Covariates in Model

Study Population* (ref., type, n)	PbB Age	Outcome Age	Mean PbB (ug/dL)	Estimated Delta IQ for PbB 5 -> 15**		SES	Smoking	Fetal Growth	Family Environment	HOME	Race	Parental IQ	Iron Status	Other
				Crude	Adj									
Cleveland (16 L 212)	Mean 0.5-3 years	4 years 10 months	9.99 at 6 months & 16.70 at both 2 years & 3 years	r=-.29	Not stated	Maternal Education	Cigarettes per Day	Birth Weight Gestation	Authoritarian Family Ideology	Total (mean of 1 2 3 and 4 years 10 months)	Child	Maternal		Child Stress Maternal Medication/Drug Use Maternal Alcohol Consumption Birth Order Child's Sex History Alcohol Abuse
Port Pirie (35 L 326)	Lifetime average - 3 years	11-13 years	Not stated	-10.2 (ln)^	-5.1 (ln)^	Daniel's Scale of Prestige of Occupations in Australia Maternal Education	Parental Smoking Habits	Birth Weight	Family Structure Family Functioning Marital Status Maternal Age Life Events	Unspecified		Maternal		Breast Feeding Feeding Method Birth Order Child's Sex Child's Age School Grade School Absence
(>= 4 years)	(>= 4 years)													
Lavrion Greece (20 X 509)	Primary school children - not specified years	Primary school children - not specified years	23.7	Not stated	-2.52^	Maternal Education Paternal Education Paternal Occupation		Birth Weight	Family Structure Marital Status Life Events			Both		Birth Order Child's Age Child's Medical History History Alcohol Abuse Father's Age Bilingualism Length of Child's Hospital Stay after Birth
Port Pirie (4 L 494)	Lifetime average - 4 years	7 years	17.6-21.5 (means of 2nd & 3rd quartiles) (GM)	Not stated	-5.5 (ln)^	Daniel's Scale of Prestige of Occupations in Australia Maternal Education Paternal Education	Parental Smoking	Birth Weight	Family Structure Maternal Age	Unspecified		Maternal		Breast Feeding Feeding Method Birth Order Child's Sex

* L=Longitudinal cohort; X=Cross-sectional.

** (ln)/(log10) = Original coefficient reported in log scale.

^ statistically significant (p < 0.05)

Covariates in Model

Study Population* (ref., type, n)	PbB Age	Outcome Age	Mean PbB (ug/dL)	Estimated Delta IQ for PbB 5 -> 15** Crude	Adj	SES	Smoking	Fetal Growth	Family Environment	HOME	Race	Parental IQ	Iron Status	Other
Port Pirie (4 L 494)	Lifetime average - 7 years	7 years	15 7-19 7 (means of 2nd & 3rd quartiles) (GM)	Not stated	-4 7 (ln)	Daniel's Scale of Prestige of Occupations in Australia Maternal Education Paternal Education	Parental Smoking	Birth Weight	Family Structure Maternal Age	Unspecified		Maternal		Breast Feeding Feeding Method Birth Order Child's Sex
Mexico City (26 X 139)	7-9 years	7-9 years	19 4	r= -24 (ln)	r=- 19 (ln)	ncome Maternal Education Paternal Education								Child's Sex Type of Housing Nutritional Status (weight for height & height for age)
Port Pirie (35 L 368)	5 years	11-13 years	14 3 (GM)	-9 2 (ln)^	-4 1 (ln)^	Daniel's Scale of Prestige of Occupations in Australia Maternal Education	Parental Smoking Habits	Birth Weight	Family Structure Family Functioning Marital Status Maternal Age Life Events	Unspecified		Maternal		Breast Feeding Feeding Method Birth Order Child's Sex Child's Age School Grade School Absence
Port Pirie (35 L 326)	Lifetime average - 11-13 years	11-13 years	14 1 (GM)	-11 9 (ln)^	-4 3 (ln)^	Daniel's Scale of Prestige of Occupations in Australia Maternal Education	Parental Smoking Habits	Birth Weight	Family Structure Family Functioning Marital Status Maternal Age Life Events	Unspecified		Maternal		Breast Feeding Feeding Method Birth Order Child's Sex Child's Age School Grade School Absence
Cincinnati (13 L 253)	Mean 39-48 months	6 5 years	14 0	-1 4	-0 2		Cigarette Consumption during Pregnancy	Birth Weight Birth Length		Unspecified		Maternal		Child's Sex
Cincinnati (13 L 253)	Mean 51-60 months	6 5 years	11 8	-2 2	-0 7		Cigarette Consumption during Pregnancy	Birth Weight Birth Length		Unspecified		Maternal		Child's Sex

* L=Longitudinal cohort, X=Cross-sectional.

** (ln)/(log10) = Original coefficient reported in log scale.

^ statistically significant (p < 0.05)

Covariates in Model

Study Population* (ref., type, n)	PbB Age	Outcome Age	Mean PbB (ug/dL)	Estimated Delta IQ for PbB 5 -> 15** Crude	Adj	SES	Smoking	Fetal Growth	Family Environment	HOME	Race	Parental IQ	Iron Status	Other
Port Pirie (35 L 360)	7 years	11-13 years	11.6 (GM)	-8.7 (ln)^	-3.1 (ln)	Daniel's Scale of Prestige of Occupations in Australia Maternal Education	Parental Smoking Habits	Birth Weight	Family Structure Family Functioning Marital Status Life Events	Unspecified		Maternal		Breast Feeding Feeding Method Birth Order Child's Sex Child's Age School Grade School Absence
Dunedin New Zealand (33 L 579)	11 years	11 years	11.1	r=-0.06 (ln)	Not stated									
Sassuolo Italy (8 L 212)	7-8 years	7-8 years	10.99 (GM)	r = -0.101 (log10)	Not stated									
San Luis Potosi Mexico (10 X 39) [reference group]	6-9 years	6-9 years	9.73 (GM)	r=.04 (ln)	r=.07 (ln)	Bronfman Index of Socioeconomic Status Maternal Education Paternal Education								Child's Sex Child's Age
San Luis Potosi Mexico (10 X 41) [exposed group]	6-9 years	6-9 years	8.98 (GM)	r=-.12 (ln)	r=-.25 (ln)	Bronfman Index of Socioeconomic Status Maternal Education Paternal Education								Child's Sex Child's Age
Port Pirie (35 L 326)	11-13 years	11-13 years	7.9 (GM)	-6.3 (ln)^	-2.6 (ln)	Daniel's Scale of Prestige of Occupations in Australia Maternal Education	Parental Smoking Habits	Birth Weight	Family Structure Family Functioning Marital Status Maternal Age Life Events	Unspecified		Maternal		Breast Feeding Feeding Method Birth Order Child's Sex Child's Age School Grade School Absence
Boston (7 L 116)	57 months	10 years	6.3	Not stated	-0.7	Hollingshead Four-Factor Index of Social Class		Birth Weight	Family Stress Marital Status Maternal Age	Scales V & V at 120 months Total at 57 months	Child	Maternal		Child Stress Birth Order Child's Sex

* L=Longitudinal cohort, X=Cross-sectional.
** (ln)/(log10) = Original coefficient reported in log scale.
^ statistically significant (p < 0.05)

Covariates in Model

Study Population* (ref., type, n)	PbB Age	Outcome Age	Mean PbB (ug/dL)	Estimated Delta IQ for PbB 5 -> 15**		SES	Smoking	Fetal Growth	Family Environment	HOME	Race	Parental IQ	Iron Status	Other
				Crude	Adj									
Boston (7 L 116)	10 years	10 years	2 9	Not stated	-5 9	Hollingshead Four-Factor Index of Social Class		Birth Weight	Family Stress Marital Status Day Care Maternal Age	Scales V & V at 120 months Total at 57 months	Child	Maternal			Child Stress Birth Order Child's Sex
Kosovo (38 L 259)	Mean AUC7 years	7 years	PbB at age 7 year's = 21 2 cumulative PbB through age 7 year's = 1 21	Not stated	-3 4 (log10)^	Maternal Education		Birth Weight	Family Structure Maternal Age	Unspecified	Child	Maternal			Child's Sex
Port Pirie (35 L 326)	Lifetime average - 5 years	11-13 years	Not stated	-10 8 (ln)^	-5 5 (ln)^	Daniel's Scale of Prestige of Occupations in Australia Maternal Education	Parental Smoking Habits	Birth Weight	Family Structure Family Functioning Marital Status Maternal Age Life Events	Unspecified		Maternal		Breast Feeding Feeding Method Birth Order Child's Sex Child's Age School Grade School Absence
Port Pirie (35 L 326)	Lifetime average - 7 years	11-13 years	Not stated	-10 5 (ln)^	-4 7 (ln)^	Daniel's Scale of Prestige of Occupations in Australia Maternal Education	Parental Smoking Habits	Birth Weight	Family Structure Family Functioning Marital Status Maternal Age Life Events	Unspecified		Maternal		Breast Feeding Feeding Method Birth Order Child's Sex Child's Age School Grade School Absence
Cincinnati (13 L 253)	Mean 66-72 months	6 5 years	Not stated	-3 3^	-1 2		Cigarette Consumption during Pregnancy	Birth Weight Birth Length		Unspecified		Maternal		Child's Sex
Cincinnati (13 L 253)	Lifetime average - 72 months	6 5 years	Not stated	-1 3	-0 1		Cigarette Consumption during Pregnancy	Birth Weight Birth Length		Unspecified		Maternal		Child's Sex

* L=Longitudinal cohort, X=Cross-sectional.

** (ln)/(log10) = Original coefficient reported in log scale.

^ statistically significant (p < 0.05)

Table 5. Studies of health endpoints other than IQ or GCI in relation to BLLs <10 μg/dL.

Study population (ref., type, n)	Health outcome	Age at measurement LL	Age at measurement Outcome	LL distribution	ovariates	Results on BLL-outcome association <10 μg/dL
HANES III (22, X, 4853)	ognitive function and academic achievement	6-16 years	6-16 years	Geometric mean (GM)=1.9 μg/dL, 98% <10 μg/dL	Sex, race/ethnicity, poverty, region, parent/caregiver education and marital status, serum ferritin, serum cotinine.	Significant inverse relationships between BLL and WRAT arithmetic, WRAT reading, WISC R block design, WISC-R digit span. For all but block design, regression slopes became more negative with restriction of analyses to children with BLL <10, <7.5, <5.0, and < 2.5 μg/dL.
Leipzig, Gardelegen, Duisberg, Germany (36, X, 384)	Attention, sensorimotor function, and cognitive function	5-7 years	5-7 years	GM=4.25 μg/dL, 95 % <9 μg/dL	For WISC vocabulary and block design: Study area, visual acuity and contrast sensitivity, parental education, sex, breastfeeding, height, nationality. For NES2 pattern comparison, pattern memory, tapping, simple reaction time, and continuous performance test: study area, visual acuity and contrast sensitivity, age, parental education, sex, birthweight, smoking in pregnancy, number of siblings, height, computer familiarity.	Significant negative association between WISC vocabulary and CPT results. Associations strongest in Gardelegen, community with lowest mean BLL.
Leipzig, Gardelegen, Duisberg, Germany (40, X, 367)	Neurobehavioral and neurophysiologic function	6 years	6 years	Median=5 μg/dL, 95 % <10 μg/dL	For NES2 tapping, benton (pattern memory), reaction time, pattern comparison: age, sex, parental education, study area. For Visual Evoked Potentials: age, gender, study area.	Significant negative association of log transformed blood lead with tapping speed and with pattern comparison. No significant association of log-transformed LL with visual evoked potential peak latencies.

Study population (ref., type, n)	Health outcome	Age at measurement		LL distribution	covariates	Results on BLL-outcome association <10 µg/dL
		LL	Outcome			
New York (24, X, 68)	Mental development	12-36 months	12-36 months	Mean=10.3 µg/dl	Receives public assistance, maternal education, HOME - Stim Q, child race, maternal IQ, anemia or low MCV, birth order, sex, age	Significantly lower Bayley MDI for children ≥10 µg/dL vs <10; scatterplot of adjusted MDI vs. BLL suggests relation linear relation continues at BLL <10.
Leipzig, Gardelegen, Duisberg, Germany (1, X, 746)	NES1 – Tapping Test and pattern recognition	5 and 6 years	5 and 6 years	Median=5 µg/dl and 95th percentile of overall frequency distribution for PbB was <10 µg/dl	Maternal education, child's sex, child's age	Authors report that after adjustment for confounders a significant deficit for tapping and pattern comparison in relation to BLL (p<0.05) was found, but no regression coefficient or dose-response analyses are presented.
Leipzig, Gardelegen, Duisberg, Germany (2, X, 384)	Visual function	5-7 years	5-7 years	GM=4.25 µg/dL, 95% <9 µg/dl	Child's age, assessment site, birth weight, child's medical history, head circumference, child weight, quality of fixation	Visual evoked potential interpeak latencies were significantly prolonged in relation to BLL for one of three visual stimuli tested and non-significantly prolonged for a second stimulus. o significant association between BLL and contrast sensitivity was seen.
Study location not stated (3, X, 400)	Neurotransmitter and neuroendocrine levels	8.5-12.3 years	8.5-12.3 years	Mean=3.95 µg/dl		No significant correlation overall between BLL and serum prolactin (Pro-S) or urinary homovanillic acid (HVA-U). Analysis performed on only those children BLL >5 µg/dL showed a weak but statistically direct relation to BLL.
NHANES III (5, X, 4391)	Stature and head circumference	1-7 years	1-7 years		Ethnic group, iron status, dietary intake, medical history, sociodemographic factors, and household characteristics	Significant inverse association of BLL to stature and head circumference (estimated decrease of 1.57 cm in stature and 0.52 cm in head circumference for each 10 µg/dL increase in BLL).

Study population (ref., type, n)	Health outcome	Age at measurement		LL distribution	covariates	Results on BLL-outcome association <10 µg/dL
		LL	Outcome			
Mexico City (30, L, 119-199)	Head circumference	every 6 months from 6-48 months	every 6 months from 6-48 months	Median postnatal varied from 7-10 µg/dl	Birth problems, child's race, maternal head circumference, head circumference at birth	Natural log of blood lead at 12, 18, and 24 months significantly related to head circumference at 36 months; Natural log of blood lead at 12 months significantly related to head circumference at 42 months. Most other partial correlations between postnatal blood were negative. Plot of covariate adjusted head circumference at 36 months vs. natural log of blood lead at 12 months shows inverse relation that appears to continue below 10 µg/dL.
Lavrion, Elefsina, Loutraki Greece (21, X, 522)	Somatic growth, including head circumference, height, and chest circumference	6-9 years	6-9 years	Mean=12.3 µg/dL, Median=9.8 µg/dl	Paternal education, paternal occupation, child's sex, child's age, iron status, assessment site, father's height, mother's height	Significant negative association of BLL and head circumference and height with scatterplot suggesting relation continues below 10 µg/dL. No significant association with chest circumference.
NHANES III (32, X, 2186) girls: 1964 with pubic hair stage, 1986 with breast development stage and 1796 with age at menarche.) (African American only)	Pubertal development in girls	8-18 years	8-18 years	GMs: Non-Hispanic whites:=1.4 µg/dL; African Americans: = 2.1; Mexican Americans= 1.7µg/dL. LLs >5 µg/dL: 2.7%, 11.6% and 12.8%, respectively	Family income, ever smoked 100 cigarettes, child's age, iron status, child's medical history, height, BMI, age squared For age at menarche: height; family income, ever smoked 100 cigarettes, child's age, iron status, child's medical history, height, MI, age squared	lood lead levels of 3 µg/dL (compared with BLLs of 1 µg/dL) were associated with significant delays in breast and pubic hair development in African American and Mexican girls. The trend was similar, but not significant, for non-Hispanic white girls. Age at menarche was also delayed in relation to higher BLLs, but the association was only significant for African-American girls.

74

Study population (ref., type, n)	Health outcome	Age at measurement LL	Age at measurement Outcome	LL distribution	ovariates	Results on BLL-outcome association <10 µg/dL
HANES III (42, X, Sample I: 1706 ages 8-16 years with pubic hair and breast development info; Sample II: 1235 girls aged 10-16 had info on menarche) (all races)	Pubertal development in girls	8-16 years	8-16 years	98.5% <10; 54.3% 0-2.0	Poverty income ratio, family size, metro residence, child's age, child's race, MI	Compared with BLLs 2.0 µg/dL and below, LLs of 2.1-4.9 were associated with significantly lower odds of attaining Tanner Stage 2 pubic hair (OR=0.48, 95% CI 0.25-0.92) and menarche (OR=0.42, 95% CI 0.18-0.97); no significant association with breast development was noted.
NHANES III (25, X, 24901)	Dental caries	2+ years	2+ years	GMs: Age 2-5 years=2.9 µg/dL, Age 6-11 years=2.1 µg/dL; Age 12+ years=2.5 µg/dL; 74-88% of participants with BLL <5 µg/dL in each age group	Poverty income ratio, maternal education, exposure to cigarette smoke, child's sex, child's age, child's race, assessment site, child's medical history, days since last dental visit, usual frequency of dental visits	omparing children 5-17 years of age in middle tertile of BLL (range of BLLs: 1.7-4.1) to those in lowest tertile, odds ratio for dental caries was 1.36 (1.01-1.83)
oston & ambridge, MA and Farmington, ME (18, X, 543)	Dental caries	6-10 years	6-10 years	Means: Cambridge/ oston 2.9 µg/dL, Farmington 1.7 µg/dL	Age, sex, family income, education of female guardian, ethnicity, maternal smoking, tooth brushing frequency, tooth brush bristle hardness, gum chewing	In Cambridge/Boston, number of carious surfaces increased significantly with log BLL in linear regression and in graph comparing children with BLL of 1, 2, and 3µg/dL. In Farmington, non-significant decrease in carious surfaces with increasing BLL
Kosovo (17, L, 281)	lood pressure	66 months	66 months	K. Mitrovica: mean=37.3 µg/dL (sd=12.0); Pristina: mean=8.7 µg/dL (sd=2.8)	For systolic blood pressure: birth order, child's sex, child's race, height, BMI For diastolic blood pressure: birth order, child's race	Figures showing adjusted mean systolic and diastolic blood pressure for 10 blood lead groups with approximately equal numbers in each ordered by blood lead shows no consistent trend among the 4 blood lead groups with BLL approximately 5-10 µg/dL.

75

Study population (ref., type, n)	Health outcome	Age at measurement LL	Age at measurement Outcome	LL distribution	covariates	Results on BLL-outcome association <10 µg/dL
elgium (29, X, 143)	Heme synthesis biomarkers	10-13 years	10-13 years	Means: Boys: <1 km: 28.7 µg/dL (SD=8); 2.5 km: 15.6 (2.9); urban: 10.6 (2.0); rural: 9.2 (2.3) Girls: <1 km: 20.7 (7.6); 2.5 km: 9.8 (3.8); urban: 9 (2.0); rural: 8.7 (17)	Not specified	Dose-effect relationships are plotted for FEP, ALAD, and ALAU. No threshold evident for ALAD inhibition. Authors state if it exists, it must be below 8-10 µg/dL. A BLL 5 threshold for increasing FEP evident at 15-20 µg/dL Pb.
oston (27, L, 249 originally recruited; 201 at 2 years)	Heme synthesis biomarkers	6-24 months	6-24 months	Mean=7 µg/dL	Not specified	No relationship between incidence of elevated erythrocyte protoporphyrin levels and BLLs below 15 µg/dL
incinnati (19, L, 165)	Heme synthesis biomarkers	6-30 months	6-30 months	Not presented	None presented, crude results only	Significant positive association reported for FEP and ZPP and log transformed BLL at all ages. Threshold for relationship at BLL between 15 and 20 µg/dL.
Pribam, Czech Republic (9, X, 246)	Renal function	12-15 years	12-15 years	Mean ranged from 8.39 µg/dL in girls in the control area to 14.9 µg/dL in boys in polluted area 2	None presented, crude results only	Urinary RBP was found to be significantly associated with BLL in a stepwise regression. When urinary RBP excretion was examined by BLL tertiles, significantly lower U-RBP was seen in the group with BLL <8.64 µg/dL compared with BLL 8.64-12.3.

Table 6. Selected methodologic details from cohort studies

Study Population	Quality Assurance Comments		
	Blood-lead Measurement	Cognitive Function Measurement	
Boston (6 7 34)	Samples were measured by capillary and venous and were analyzed by ASV and GFAAS Blood specimens for 6- 12- 18 and 24-month specimens were collected in capillary tubes by trained technicians Blood samples were assayed in duplicate or triplicate The analytical system was calibrated with aqueous standards of known lead concentrations Each batch of samples was accompanied by a blood sample of known lead concentrations to quantify intralaboratory reliability Several standardized blood samples with lead concentrations also were included after they became available in 1982 from CDC (Rabinowitz et al 1985) 57-month venous blood samples were obtained Lead was measured in duplicate by GFAAS An aliquot of a standardized blood sample provided by the National Bureau of Standards was included in each batch of samples (Bellinger et al 1991)	MD was administered at 6-month intervals beginning at 6 months of age by examiners blind to the infants' lead levels (Bellinger et al 1985) For W SC-R most children were tested in a single session 2 were seen in a second session to complete testing and 7 were tested in their homes by parental request Psychologists were blind to all aspects of child's developmental and lead exposure histories	
Cincinnati (13)	Samples were measured by venipuncture heel stick and finger stick for infants and were analyzed by ASV Blood samples were obtained using either venipuncture or heel stick Approximately 72% of all samples are venipuncture For heel stick two capillary tubes were filled for duplicate PbB determination f venipuncture was possible pediatric vacutainer tubes were filled one for PbB determination and a second for serum iron and total iron binding capacity (T BC) analyses The sample was aliquoted and duplicate analyses performed according to a predetermined protocol using ASV The laboratory participates in both the CDC and PA State Blood Lead and Protoporphyrin Programs A series of bench-top QC samples and blind QC samples were analyzed with each run (Bornschein et al 1985)	For W SC-R one experienced psychometrician performed all the assessments Children were tested at a pediatric clinic The examiner was blind to the exposure levels of the child For MD all assessments took place in a prenatal and child welfare clinic Psychometric tests were administered at an inner-city health clinic by the study leader or trained assistant with whom inter-tester reliability had been previously established Testers were blind to children's blood-lead levels	
Cleveland (14 15 16)	Samples were measured by venous and were analyzed by GFAAS Blood samples were collected in heparinized plastic syringes which had been determined to be free of trace metals The concentration of lead in whole blood samples was determined by GFAAS All samples were run in duplicate The obtained within-run (same day) reproducibility was evaluated for a sample of adult whole blood The obtained values were 55 2 ug/dl 1 34 and 2 4% respectively for the mean SD and coefficient of variation Regular assessment of accuracy and precision using CDC samples of bovine blood were conducted and found to be within the certified range Two inter-laboratory reviews were conducted for further determination of accuracy Blood-lead levels were not adjusted for hematocrit (Ernhart et al 1985)	WPPS MD and Stanford Binet Q tests were conducted by well-trained examiners blind to all risk and background information Home testing was used to control attrition to minimize bias in attrition and to facilitate administration of the HOME nventory nter-observer agreement was checked through observation and duplicate scoring by a supervisor for approximately one out of every 26 examinations Agreement was maintained at r= 99 Answer sheets were checked for possible irregularities by the supervisor within a few days of each administration	
Costa Rica (41)	Samples were measured by venous and were analyzed by GFAAS Venipuncture samples were taken and red blood cells were promptly separated and frozen for future analysis in the U S The frozen red cells were analyzed using GFAAS in a laboratory that participates in CDC's Maternal and Child Health Resources Development Proficiency Testing Program for Blood Lead Quality control was monitored through certified controls obtained from the National Bureau of Standards Red cell lead values were converted to whole blood-lead leve s using the formula of Rosen et al (1974)	Spanish versions of Bayley MD and WPPS were used in the assessment A single tester trained by one of the primary investigators and the most senior research psychologist in the country administered the assessments The tester was blind to the children's iron status and never knew the blood-lead levels (these were performed in the U S) (Lozoff personal communication)	
Kosovo (37 38)	Samples were measured by venous and were analyzed by GFAAS All blood specimens were refrigerated on site and transported on wet ice to Columbia University where all assays were performed The laboratory participates in CDC's PBB QC program and is certified by OSHA Over the study period interclass correlation with QC values was computed with correlation coefficients of 95 for PbB	Three Yugoslavian psychologists scored the W SC-R and the McCarthy GC independently All interviews and assessment instruments were translated and administered in the two dominant languages of the region Serbo-Croatian and Albanian Training and reliability visits occurred The average interclass correlation for 96 tests over study period was calculated	
Mexico City (31)	Samples were measured by venous and were analyzed by ASV Samples were analyzed at Environmental Sciences Associates (ESA) Laboratories nc which is a CDC reference lab for the Blood Lead Proficiency Testing Program and also participates in the New York State Department of Control Program All samples were analyzed using ASV Samples with mean duplicate values < 5 ug/dl were reanalyzed in duplicate by graphite furnace AAS Mean values of the duplicates were used as data (Rothenberg et al 1994)	Four trained psychologists blind to children's lead levels administered the McCarthy GC As there were no norms for the McCarthy scale in the Mexican population the U S norms were used to calculate GC with a Spanish translation of the test nterexaminer reliability was assessed by calculating the correlation in GC scores assigned by two of the psychologists with the scores of a third psychologist whom they observed applying the test in all possible combinations with 10 subjects for each combination Mean observer-examiner correlation was 99	

APPENDIX A:

LITERATURE REVIEW AND CLASSIFICATION UPDATE

The literature review began with the Agency for Toxic Substances and Disease Registry's Toxicological Profile for Lead (ATSDR Tox Profile), published July 1999. The Health Effects chapter was thoroughly read and all articles relating to low blood lead levels in children were chosen, regardless of whether they demonstrated significant results. New literature searches were then performed by Battelle's Technical Information Center. The year 1995 was chosen as the cutoff date for the new searches because the WG felt that, before this time, research rarely focused on BLLs <10 µg/dL, and that most relevant articles before 1995 were cited in the ATSDR Toxicological Profile. Searches were performed on a variety of databases using DIALOG and a set of keywords.

The following is an example of the DIALOG, including databases and keywords:

SYSTEM:OS - DIALOG OneSearch
File   6:NTIS 1964-2003/May W3 (c) 2003 NTIS, Intl Cpyrght All Rights Res
File 103:Energy SciTec 1974-2003/May B1 (c) 2003 Contains copyrighted material
File 266:FEDRIP 2003/Mar Comp & dist by NTIS, Intl Copyright All Rights Res
File 161:Occ.Saf.& Hth. 1973-1998/Q3 (c) Format only 1998 The Dialog Corp.
File 156:ToxFile 1965-2003/May W2 (c) format only 2003 The Dialog Corporation
File 155:MEDLINE(R) 1966-2003/May W2 (c) format only 2003 The Dialog Corp.
File 162:Global Health 1983-2003/Apr (c) 2003 CAB International
File   71:ELSEVIER BIOBASE 1994-2003/May W3 (c) 2003 Elsevier Science B.V.
File   40:Enviroline(R) 1975-2003/May
File   73:EMBASE 1974-2003/May W1 (c) 2003 Elsevier Science B.V.
File   34:SciSearch(R) Cited Ref Sci 1990-2003/May W2 (c) 2003 Inst for Sci Info
File   5:Biosis Previews(R) 1969-2003/May W2 (c) 2003 BIOSIS

Set Items Description
S1   512735 NATAL? OR PRENATAL? OR PERINATAL? OR POSTNATAL?
S2 1244432 INFANT? ? OR INFANCY
S3 2607491 CHILD? ? OR CHILDREN? ?
S4   253558 LEAD/TI,DE,ID
S5   184354 PB
S6   68959 RN=7439-92-1
S7 5798237 BLOOD
S8   14048 (S1:S3) AND (S4:S6) AND S7
S9 2153692 GROWTH/TI,DE,ID
S10   31450 STATURE
S11   634981 NUTRITION
S12   169948 HEARING
S13   200409 (RENAL OR KIDNEY)(3N)FUNCTION?
S14   669012 BLOOD()PRESSURE
S15   13 HEMESYNTHESIS
S16   61334 HEMATOPOIESIS
S17   20269 (VITAMIN()D)(3N)METABOLI?
S18   1441 S8 AND (S9:S17)
S19   438 S18 AND PY=1990:1996
S20   422 S19/ENG OR (S19 AND LA=ENGLISH)
S21   353 S20/HUMAN
S22   190 RD (unique items)
S23   190 Sort S22/ALL/PY,D

```
S24      19583  NEUROBEHAVIO?
S25     272980  NEUROLOGICAL?
S26     166224  NEUROLOGIC
S27     148942  NEUROTOXIC?
S28      15459  NEURODEVELOPMENT?
S29       7061  COGNITIVE()DEVELOPMENT
S30    2878983  BEHAVIOR? OR BEHAVIOUR?
S31          3  IMPULSITIVITY
S32      54987  HYPERACTIVITY
S33      10725  ADHD
S34      31451  IQ OR (INTELLIGENCE()QUOTIENT? ?)
S35       3350  WISC
Ref  Items Index-term
E1     715223  *DC=A8.186.          (Central nervous system)
E2     645734  DC=A8.186.211.      (Brain)
E3      23250  DC=A8.186.211.132.          (Brain stem)
S36    715223  DC='A8.186.':DC='A8.186.211.132.'
S37       2936  (S8 AND (S24:S36)) NOT S18
S38        964  S37 AND PY=1990:1996
S39        941  S38/ENG OR (S38 AND LA=ENGLISH)
S40        808  S39/HUMAN
S41        415  RD (unique items)
```

This literature search was first run for the years 1995-2002. In spring, 2003, the search was rerun for the years 2002-2003 to determine the relevance of recently published articles. Also at this time, the search was rerun for the years 1990-1996 for relevant articles that were not cited in the ATSDR Toxicological Profile. Titles and abstracts from each literature search were reviewed, and relevant articles were ordered for further review. Additional articles were identified while reviewing the selected articles and were added to the list of references, as were several articles recommended by workgroup members.

The table below provides a summary of all the articles obtained from the various sources. This table shows when the search was performed, the years covered in the search, the number of articles found in the literature search, the number of articles ordered after the titles and abstracts had been reviewed, and the number of articles that were relevant for abstraction.

Summary of Literature Review Results

Date of Search	Years Covered	Number of Unique References Found	Number of reviewed for relevance	Number of Articles Abstracted into the Database
9/02	1995-2002	327	79	12
4/03	2002-2003	119	14	4[a]
5/03	1990-1996	605	25	4
ATSDR Tox Profile	Prior to 1996	-	107	24
Referrals[b]	various	10	12	6
		Total:	**235**	**50**
	Relevant articles cited in Tables 2- 5			**42**

[a]A 5th article, Stone et al. 2003, was obtained from this search and is not abstracted into the database but its relevance is discussed elsewhere in the report.

[b]Referrals include articles that were recommended by workgroup members, as well as those articles cited as references in studies identified in the ATSDR Tox Profile or literature searches.

APPENDIX B:

DISCUSSION OF CRITIQUE OF
NHANES III DATA BY STONE ET AL. (2003)

Stone et al. (2003) reanalyzed the data used by Lanphear et al. (2000). While the results they present are largely consistent with the findings of Lanphear et al. (2000), they provided a critique of the validity of the NHANES III data for evaluating lead-related impacts on neuropsychological development in children. Because their critique cuts across on neuropyschological measurements performed in the survey, the main points of their paper are summarized in this appendix, as follows.

- Stone et al. note that the weighted mean values for the four measures used by Lanphear et al. are below the predicted mean based on standardization data for these tests collected in the early 1970s for the WISC-R) and early 1980s for the WRAT. Stone et al. argue that the mean values should be higher than predicted by the standardization means because of secular improvements in cognitive test scores. One possible reason cited for the discrepancy is that NHANES tests were not administered by a psychologist. It is unclear, however, if the population sample used in the standardization data was equally representative of the U.S. population at that time or if changes in the population composition since then would lead to an increase or decrease in overall mean test performance. More importantly, it is unclear how a bias in mean score, even if real, and the use of non-psychologists for testing could produce associations between BLLs and test scores, given that examiners could not have known the BLLs of participants. If non-psychologists produced less precise test results than psychologists would have, the expected impact on regression coefficients would be a bias toward the null.
- The age-adjusted scores used in NHANES are correlated with age, and they should not be. Stone et al. show that age is negatively correlated with arithmetic, block design, and digit span and positively correlated with reading. However, since BLLs decrease with age across the age range studied, the negative correlations would tend to produce a trend towards higher scores with increasing blood lead for those tests, the opposite of the findings of Stone et al.
- Imputation of missing covariate values was performed for a substantial proportion of observations in the analyses performed by Lanphear et al. While imputation could increase covariate mismeasurement and residual confounding, analyses presented by Stone et al. demonstrate essentially similar findings when analyses are restricted to observations with full rank data.
- Relevant covariates, including whether a child has repeated a grade, whether interviews were in Spanish, and several other factors, were not included in analyses. However, two problems are evident in alternative "two stage" analysis provided by Stone et al. First, it uses predicted rather than residual blood lead level as an independent variable in a model relating blood lead to test scores. This amounts to testing the relation to test scores of a linear combination of covariates, many included in the model with test score as the outcome. In addition at least one variable – having to repeat a grade – is included as a covariate, possibly result serious over control as

discussed earlier. Lead associated cognitive and behavioral effects have, not surprisingly, been associated with an increased risk of failure to complete high school. Thus, controlling for failure to complete a grade could amount to controlling for an effect of, rather than a confounder of the lead effect.

As a whole, the Stone et al. critique of the NHANES III data do not provide a convincing argument that the findings reported by Lanphear et al. (2000) result from problems with the sample or testing methods. However, the WG did consider the limitations of the Lanphear et al. study, including its cross-sectional design and limited data on potential confounders. This study was weighed in the overall context of other relevant studies, including the more persuasive cohort studies, which are largely consistent with the associations Lanphear et al. report.

REFERENCES

Note: Table R-1, which follows this alphabetical listing of references, identifies the numbered references that are used in Tables 2 through 6 (which also are included in the alphabetical listing).

Agency for Toxic Substances and Disease Registry. Toxicological profile for lead. Atlanta: U.S. Department of Health and Human Services, 1999.

Al-Saleh, I., Nester, M., DeVol, E., Shinwari, N., Munchari, L., Al-Shahria, S. Relationships between blood lead concentrations, intelligence, and academic achievement of Saudi Arabian schoolgirls. Int J Hyg Environ Health. 2001;204:165–174.

Altmann, L., Sveinsson, K., Kramer, U., Weishoff-Houben, M., Turfeld, M., Winneke, G., et al. Visual functions in 6-year old children in relation to lead and mercury levels. Neurotoxicol Teratol. 1998;20:9–17.

Altmann, L., Sveinsson, K., Kramer, U., Winneke, G., Wiegand, H. Assessment of neurophysiologic and neurobehavioral effects of environmental pollutants in 5- and 6-year-old children. Environ Res. 1997;73:125–131.

Alvarez Leite, E.M., Leroyer, A., Nisse, C., Haguenoer, J.M., de Burbure, C.Y., Buchet, J.P., et al. Urinary homovanillic acid and serum prolactin levels in children with low environmental exposure to lead. Biomarkers. 2002;7:49–57.

Anderson, S.E., Dallal, G.E., Must, A. Relative weight and race influence average age at menarche: results from two nationally representative surveys of U.S. girls studied 25 years apart. Pediatrics. 2003;111:844–850.

Annest, J.L., Pirkle, J.L., Makuc, D., Neese, J.W., Bayse, D.D., Kovar, M.G. Chronological trend in blood lead levels between 1976 and 1980. N Engl J Med. 1983;308:1373–1377.

Baghurst, P.A., McMichael, A.J., Wigg, N.R., Vimpani, G.V., Robertson, E.F., Roberts, R.J., et al. Environmental exposure to lead and children's intelligence at the age of seven years. The Port Pirie Cohort Study. N Engl J Med. 1992;327:1279–1284.

Baghurst, P.A., Tong, S.L., Sawyer, M.G., Burns, J.M., McMichael, A.J. Sociodemographic and behavioural determinants of blood lead concentrations in children aged 11-13 years. The Port Pirie Cohort Study. Med J Aust. 1999;170:63–67.

Bailey, D.A., Martin, A.D., McKay, H.A., Whiting, S., Mirwald, R. Calcium accretion in girls and boys during puberty: a longitudinal analysis. J Bone Miner Res. 2000;15:2245–2250.

Ballew, C., Khan, L.K., Kaufmann, R., Mokdad, A., Miller, D.T., Gunter, E.W. Blood lead concentration and children's anthropometric dimensions in the third National Health and Nutrition Examination Survey (NHANES III), 1988–1994. J Pediatr. 1999;134:623–630.

Bannon, D.I., Chisolm, J.J. Anodic stripping voltammetry compared with graphite furnace atomic absorption spectrophotometry for blood lead analysis. Clin Chem. 2001;47:1703–1704.

Bellinger, D.C. Interpreting the literature on lead and child development: the neglected role of the "experimental system." Neurotoxicol Teratol. 1995;17:201–212.

Bellinger, D.C., Leviton, A., Rabinowitz, M.B., Needleman, H.L., Waternaux, C. Correlates of low-level lead exposure in urban children at 2 years of age. Pediatrics. 1986;77:826–833.

Bellinger, D.C., Leviton, A., Waternaux, C. Lead, IQ and social class. Int J Epidemiol. 1989;18:180–185.

Bellinger, D.C., Leviton, A., Waternaux, C., Allred, E. Methodological issues in modeling the relationship between low-level lead exposure and infant development: examples from the Boston Lead Study. Environ Res. 1985;38:119–129.

Bellinger, D.C., Leviton, A., Waternaux, C., Needleman, H.L., Rabinowitz, M.B. Longitudinal analyses of prenatal and postnatal lead exposure and early cognitive development. N Engl J Med. 1987;316:1037–1043.

Bellinger, D.C., Leviton, A., Waternaux, C., Needleman, H.L., Rabinowitz, M.B. Low-level lead exposure and early development in socioeconomically advantaged urban infants. In: Smith M, Grant L.D.; Sors A Ieds. Lead Exposure and Child Development: An International Assessment. Lancaster, UK: Kluwer Publishers. 1989:345–356.

Bellinger, D.C., Leviton, A., Waternaux, C., Needleman, H.L., Rabinowitz, M.B. Low-level lead exposure, social class, and infant development. Neurotoxicol Teratol. 1988;10:497–503.

Bellinger, D.C., Needleman, H.L. Correspondence: Intellectual Impairment and blood lead levels. N Engl J Med. 2003;349:500–502.

Bellinger, D.C., Sloman, J., Leviton, A., Rabinowitz, M.B., Needleman, H.L., Waternaux, C. Low-level lead exposure and children's cognitive function in the preschool years. Pediatrics. 1991;87:219–227.

Bellinger, D.C., Stiles, K.M., Needleman, H.L. Low-level lead exposure, intelligence and academic achievement: a long-term follow-up study. Pediatrics. 1992;90:855–861.

Bergomi, M., Borella, P., Fantuzzi, G., Vivoli, G., Sturloni, N., Cavazzuti, G., et al. Relationship between lead exposure indicators and neuropsychological performance in children. Dev Med Child Neurol. 1989;31:181–190.

Bernard, A., Vyskocil, A., Roels, H.A., Kriz, J., Kodl, M., Lauwerys, R. Renal effects in children living in the vicinity of a lead smelter. Environ Res. 1995;68:91–95.

Bornschein, R.L., Succop, P.A., Dietrich, K.N., Clark, S.C., Que Hee, S., Hammond, P.B. The influence of social and environmental factors on dust lead, hand lead, and blood lead levels in young children. Environ Res. 1985;38:108–118.

Brandt, J., Van Gorp, W. American Academy of Clinical Neuropsychology policy on the use of non-doctoral level personnel in conducting clinical neuropsychological evaluations. J Clin Exp Neuropsychol. 1999;21:1–1.

Brockel, B.J., Cory-Slechta, D.A. Lead, attention, and impulsive behavior: changes in a fixed-ratio waiting-for-reward paradigm. Pharmacol Biochem Behav. 1998;60:545–552.

Bronner, F., Abrams, S.A. Development and regulation of calcium metabolism in healthy girls. J Nutr. 1998;128:1474–1480.

Bushnell, P.J., Bowman, R.E. Persistence of impaired reversal learning in young monkeys exposed to low levels of dietary lead. J Toxicol Environ Health. 1979;5:1015–1023.

Cadogan, J., Blumsohn, A., Barker, M.E., Eastell, R. A longitudinal study of bone gain in pubertal girls: anthropometric and biochemical correlates. J Bone Miner Res. 1998;13:1602–1612.

Cake, K.M., Bowins, R.J., Vaillancourt, C., Gordon, C.L., MacNutt, R.H., Laporte, R., et al. Partition of circulating lead between serum and red cells is different for niternal and external sources of exposure. Am J Ind Med. 1996;29:440–445.

Calderon, J., Navarro, M.E., Jimenez-Capdeville, M.E., Santos-Diaz, M.A., Golden, A., Rodriguez-Leyva, I., et al. Exposure to arsenic and lead and neuropsychological development in Mexican children. Environ Res. 2001;85:69–76.

Canfield, R.L., Henderson, C.R., Cory-Slechta, D.A., Cox, C., Jusko, T.A., Lanphear, B.P. Intellectual impairment in children with blood lead concentrations below 10 µg per deciliter. N Engl J Med. 2003;348:1517–1526.

Centers for Disease Control and Prevention. Managing Elevated Blood Lead Levels Among Young Children: Recommendations from the Advisory Committee on Childhood Lead Poisoning Prevention. Nutritional Interventions. Atlanta: U.S. Department of Health and Human Services, 2002. Available at http://www.cdc.gov/nceh/lead/CaseManagement/caseManage_chap4.htm. Accessed February 23, 2004.

Centers for Disease Control and Prevention. Second National Report on Human Exposure to Environmental Chemicals. Atlanta: U.S. Department of Health and Human Services, 2003. NCEH Pub. No. 02-0716. Available at http://www.cdc.gov/exposurereport/2nd/pdf/secondner.pdf. Accessed February 23, 2004.

Centers for Disease Control and Prevention. Preventing Lead Poisoning in Young Children. Atlanta: U.S. Department of Health and Human Services, 1991. Available at http://www.cdc.gov/nceh/lead/publications/books/plpyc/contents.htm. Accessed February 23, 2004.

Charney, E., Sayre, J., Coulter, M. Increased lead absorption in inner city children: Where does the lead come from? Pediatrics. 1980;65:226–231.

Cheng, Y., Schwartz, J., Sparrow, D., Aro, A., Weiss, S., Hu, H. Bone lead and blood lead levels in relation to baseline blood pressure and the prospective development of hypertension: The Normative Aging Study. Am J Epidemiol. 2001;153:164–171.

Cohen, D.J., Johnson, W.T., Caparulo, B.K. Pica and elevated blood lead level in autistic and atypical children. Am J Dis Child. 1976;130:47–48.

Cooney, G.H., Bell, A., McBride, W., Carter, C. Low-level exposures to lead: the Sydney lead study. Dev Med Child Neurol. 1989;31:640–649.

Cory-Slechta, D.A., Weiss, B., Cox, C. Performance and exposure indices of rats exposed to low concentrations of lead. Toxicol Appl Pharmacol. 1985;78:291–299.

Davis, J.M., Otto, D.A., Weil, D.E., Grant, L.D. The comparative developmental neurotoxicity of lead in humans and animals. Neurotoxicol Teratol. 1990;12:215–229.

Deng, W., Poretz, R.D. Protein kinzse C activation is required for the lead-induced inhibition of proliferation and differentiation of cultured oligodendrglial progenitor cells. Brain Res. 2002;929:87–95.

Dietrich, K.N., Berger, O.G., Succop, P.A., Hammond, P.B., Bornschein, R.L. The developmental consequences of low to moderate prenatal and postnatal lead exposure: intellectual attainment in the Cincinnati Lead Study Cohort following school entry. Neurotoxicol Teratol. 1993;15:37–44.

Dietrich, K.N., Ris, M.D., Succop, P.A., Berger, O.G., Bornschein, R.L. Early exposure to lead and juvenile delinquency. Neurotoxicol Teratol. 2001;23:511–518.

Dietrich, K.N., Succop, P.A., Bornschein, R.L., Krafft, K.M., Berger, O.G., Hammond, P.B., et al. Lead exposure and neurobehavioral development in later infancy. Environ Health Perspect. 1990;89:13–19.

Dresner, D.L., Ibrahim, N.G., Mascarenhas, B.R., Levere, R.D. Modulation of bone marrow heme and protein synthesis by trace elements. Environ Res. 1982;28:55–66.

Ernhart, C.B., Morrow-Tlucak, M., Marler, M.R., Wolf, A.W. Low level lead exposure in the prenatal and early preschool periods: early preschool development. Neurotoxicol Teratol. 1987;9:259–270.

Ernhart, C.B., Morrow-Tlucak, M., Wolf, A.W. Low level lead exposure and intelligence in the preschool years. Sci Total Environ. 1988;71:453–459.

Ernhart, C.B., Morrow-Tlucak, M., Wolf, A.W., Super, D., Drotar, D. Low level lead exposure in the prenatal and early preschool periods: intelligence prior to school entry. Neurotoxicol Teratol. 1989;11:161–170.

Ernhart, C.B., Wolf, A.W., Sokol, R.J., Brittenham, G.M., Erhard, P. Fetal lead exposure: antenatal factors. Environ Res. 1985;38:54–66.

Eskenazi, B., Castorina, R. Association of prenatal maternal or postnatal child environmental tobacco smoke exposure and neurodevelopmental and behavioral problems in children. Environ Health Perspect. 1999;107:991–1000.

Factor-Litvak, P., Kline, J.K., Popovac, D., Hadzialjevic, S., Lekic, V., Preteni-Rexhepi, E., et al. Blood lead and blood pressure in young children. Epidemiology. 1996;7:633–637.

Fowler, B.A., Kimmel, C.A., Woods, J.S., McConnell, E.E., Grant, L.D. Chronic low-level lead toxicity in the rat: III. An integrated assessment of long-term toxicity with special reference to the kidney. Toxicol Appl Pharmacol. 1980;56:59–77.

Fulton, M., Raab, G., Thomson, G., Laxen, D., Hunter, J., Hepburn, W. Influence of blood lead on the ability and attainment of children in Edinburgh. Lancet. 1987;1:1221–1226.

Garrido Latorre, E., Hernandez-Avila, M., Tamayo Orozco, J., Albores Medina, C.A., Aro, A., Palazuelos, E., et al. Relationship of blood and bone lead to menopause and bone mineral density among middle-age women in Mexico City. Environ Health Perspect. 2002;110:A625–A630.

Gemmel, A., Tavares, M., Alperin, S., Soncini, J., Daniel, D., Dunn, J., et al. Blood lead level and dental caries in school-age children. Environ Health Perspect. 2002;110:A625–A630.

Gilbert, S.G., Rice, D.C. Low-level lifetime exposure produces behavioral toxicity (spatial discrimination reversal) in adult monkeys. Toxicol Appl Pharmacol. 1987;91:484–490.

Goldstein, G.W. Evidence that lead acts as a calcium substitute in second messenger metabolism. Neurotoxicology. 1993;14:97–101.

Grandjean, P., Weihe, P., White, R.F., Debes, F., Araki, S., Yokoyama, K., et al. Cognitive deficit in 7-year-old children with prenatal exposure to methylmercury. Neurotoxicol Teratol. 1997;19:417–428.

Grantham-McGregor, S., Ani, C. A review of studies on the effect of iron deficiency on cognitive development in children. J Nutr. 2001;131:649s–668s.

Greenland, S., Robins, J. M. Confounding and misclassification. Am J Epidemiol. 1985;122:495–506.

Grosse, S.D., Matte, T.D., Schwartz, J., Jackson, R.J. Economic gains resulting from the reduction of children's exposure to lead in the United States. Environ Health Perspect. 2002;110:563–569.

Hammond, P.B., Bornschein, R.L., Succop, P.A. Dose-effect and dose-response relationships of blood lead to erthryocytic protoporphyrin in young children. Environ Res. 1985;38:187–196.

Hatzakis, A., Kokkevi, A., Katsouyanni, K., Maravelias, K., Salaminios, F., Kalandi, A., et al. Psychometric intelligence and attentional performance deficits in lead-exposed children. In: Lindberg S.E., Hutchinson T.C., eds. International Conference on Heavy Metals in the Environment, Vol. 1. Edinburgh, UK: CEP Consultants, 1987;204–209.

Hayes, E.B., McElvaine, M.D., Orbach, H.G., Fernandez, A.M., Lyne, S., Matte, T.D. Long-term trends in blood lead levels among children in Chicago: relationship to air lead levels. Pediatrics. 1994;93:195–200.

Hernandez-Avila, M., Smith, D., Meneses, F., Sanin, L. H., Hu, H. The influence of bone and blood lead on plasma lead levels in environmentally exposed adults. Environ Health Perspect. 1998;106:473–477.

Hussain, R.J., Parsons, P.J., Carpenter, D.O. Effects of lead on long-term potentiation in hippocampal CA3 vary with age. Brain Res Dev Brain Res. 2000;121:243–252.

Jacobs, D.E., Clickner, R.P., Zhou, J.D., Viet, S.M., Marker, D.A., Rogers, J.W., et al. The prevalence of lead-based paint hazards in U.S. housing. Environ Health Perspect. 2002;110:A599–A606.

Juberg, D.R., Alfano, K., Coughlin, R.J., Thompson, K.M. An observational study of object mouthing behavior by young children. Pediatrics. 2001;107:135–142.

Kafourou, A., Touloumi, G., Makropoulos, V., Loutradi, A., Papanagioutou, A., Hatzakis, A. Effects of lead on the somatic growth of children. Archives Env Health. 1997;52:377–383.

Kaufman, A.S. Do low levels of lead produce IQ loss in children? A careful examination of the literature. Archives of Clinical Neuropsychology. 2001;16:303–341.

Kusell, M., Lake, L., Anderson, L.A., Gerschenson, L.E. Cellular and molecular toxicology of lead. II. Effect of lead on d-aminolevulinic acid synthetase of cultured cells. J Toxicol Environ Health. 1978;4:515–525.

Lanphear, B.P., Byrd, R. S., Auinger, P., Schaffer, S.J. Community characteristics associated with elevated blood lead levels in children. Pediatrics. 1998;101:264–271.

Lanphear, B.P., Dietrich, K.N., Auinger, P., Cox, C. Cognitive deficits associated with blood lead concentrations <10 ug/dL in US children and adolescents. Public Health Rep. 2000;115:521–529.

Lanphear B.P., Hornung R., Khoury J., Yolton K., Baghurst P., Bellinger D.C., et al. Low-level environmental lead exposure and children's intellectual function: an international pooled analysis. Environ Health Perspect. 2005;113:894–899.

Leggett, R.W. An age-specific kinetic model of lead metabolism in humans. Environ Health Perspect. 1993;101:598–616.

Levin, E.D., Bowman, R.E. Long-term lead effects on the Hamilton Search Task and delayed alternation in monkeys. Neurobehav Toxicol Teratol. 1986;8:219–224.

Lozoff, B., Jimenez, E., Wolf, A.W. Long-term developmental outcome of infants with iron deficiency. N Engl J Med. 1991;325:687–694.

Mahaffey, K.R., Rosen, J.F., Chesney, R.W., Peeler, J.T., Smith, C.M., DeLuca, H.F. Association between age, blood lead concentration, and serum 1,25-dihydroxycholecalciferol levels in children. N Engl J Med. 1982;307:573–579.

Mannino, D.M., Albalak, R., Grosse, S.D., Repace, J. Second-hand smoke exposure and blood lead levels in U.S. children. Epidemiology. 2003;14:719–727.

Markovac, J., Goldstein, G.W. Picomolar concentrations of lead stimulate brain protein kinase C. Nature. 1988;334:71–73.

Matthee, A., von Schirnding, Y.E., Levin, J., Ismail, A., Huntley, R., Cantrell, A. A survey of blood lead levels among young Johannesburg school children. Environ Res. 2002;90:181–184.

McElvaine, M.D., DeUngria, E.G., Matte, T.D., Copley, C.G., Binder, S. Prevalence of radiographic evidence of paint chip ingestion among children with moderate to severe lead poisoning; St. Louis, Missouri, 1989 through 1990. Pediatrics. 1992;89:740–742.

McMichael, A.J., Baghurst, P.A., Wigg, N. R., Vimpani, G. V., Robertson, E. F., Roberts, R. J. Port Pirie Cohort Study: environmental exposure to lead and children's abilities at the age of four years. N Engl J Med. 1988;319:468–475.

Mendelsohn, A. L., Dreyer, B. P., Fierman, A. H., Rosen, C. M., Legano, L. A., Kruger, H. A., Lim, S. W., Barasch, S., Au, L., Courtlandt, C. D. Low-level lead exposure and cognitive development in early childhood. J Dev Behav Pediatr. 1999;20:425–431.

Meyer, P.A., Pivetz, T., Dignam, T.A., Homa, D.M., Schoonover, J, Brody, D. Surveillance for elevated blood lead levels among children—United States, 1997–2001. In: CDC Surveillance Summaries. MMWR 2003;52(No. SS-10):1–21.

Moss, M.E., Lanphear, B.P., Auinger, P. Association of dental caries and blood lead levels. J Am Med Assoc. 1999;281(24):2294–2298.

Munoz, H., Meneses-Gonzalez, F., Romieu, I., Hernandez-Avila, M., Palazuelos, E., Mancilla-Sanchez, T. Blood lead level and neurobehavioral development among children living in Mexico City. Arch Environ Health. 1993;48:132–138.

Parsons, P.J., Reilly, A.A., Esernio-Jenssen, D. Screening children exposed to lead: an assessment of the capillary blood lead fingerstick test. Clin Chem. 1997;43:302–311.

Parsons, P.J., Slavin, W. A rapid Zeeman graphite furnace atomic absorption spectrometric method for the determination of lead in blood. Spetrochim Acta B. 1993;48:925–939.

Paschal, D.C., Caldwell, K.L., Ting, B.G. Determination of lead in whole blood using inductively coupled argon plasma mass spectrometry with isotope dilution. J Anal At Spectrom. 1995;10:367–370.

Pirkle, J.L., Brody, D., Gunter, E.W., Kramer, R.A., Paschal, D.C., Flegal, K.M., et al. The decline in blood lead levels in the United States—The National Health and Nutrition Examination Surveys (NHANES). JAMA 1994;272:284–291.

Pirkle, J.L., Kaufmann, R., Brody, D., Hickman, T., Gunter, E.W., Paschal, D.C. Exposure of the U.S. population to lead, 1991–1994. Environ Health Perspect. 1998;106:745–750.

Rabinowitz, M.B., Leviton, A., Needleman, H.L. Occurrence of elevated protoporphyrin levels in relation to lead burden in infants. Environ Res. 1986;39:253–257.

Rabinowitz, M.B., Leviton, A., Needleman, H.L. Variability of blood lead concentrations during infancy. Arch Environ Health. 1984;39:74–77.

Rabinowitz, M.B., Leviton, A., Needleman, H.L., Bellinger, D.C., Waternaux, C. Environmental correlates of infant blood lead levels in Boston. Environ Res. 1985;38:96–107.

Rahman, A., Maqbool, E., Zuberi, H.S. Lead-associated deficits in stature, mental ability and behaviour in children in Karachi. Ann Trop Paediatr. 2002;22:301–311.

Rice, D.C. Chronic low-level lead exposure from birth produces deficits in discrimination reversal in monkeys. Toxicol Appl Pharmacol. 1985;77:201–210.

Roda, S.M., Greenland, R.D., Bornschein, R.L., Hammond, P.B. Anodic stripping voltammetry procedure modified for improved accuacy of blood lead analysis. Clin Chem. 1988;34:563–567.

Roels, H.A., Lauwerys, R. Evaluation of dose-effect and dose-response relationships for lead exposure in different Belgian population groups (fetus, child, adult men and women). Trace Elem Med. 1987;4:80–87.

Rosen, J.F., Zarate-Salvador, C., Trinidad, E.E. Plasma lead levels in normal and lead-intoxicated children. Pediatrics. 1974;84:45–48.

Rothenberg, S.J., Karchmer, S., Schnaas, L., Perroni, E., Zea, F., Alba, J.F. Changes in serial blood lead levels during pregancy. Environ Health Perspect. 1994;102:876–880.

Rothenberg, S.J., Schnaas, L., Perroni, E., Hernandez, R.M., Martinez, S., Hernandez, C. Pre- and postnatal lead effect on head circumference: a case for critical periods. Neurotoxicol Teratol. 1999;21:1–11.

Savitz, D.A., Baron, A.E. Estimating and correcting for confounder misclassifcation. Am J Epidemiol. 1989;129:1062–1071.

Schlenker, T.L., Fritz, C.J., Mark, D., Layde, M., Linke, G., Murphy, A., et al. Screening for pediatric lead poisoning. Comparability of simultaeously drawn capillary and venous blood samples. JAMA. 1994;271:1346–1348.

Schnaas, L., Rothenberg, S.J., Perroni, E., Martinez, S., Hernandez, C., Hernandez, R.M. Temporal pattern in the effect of postnatal blood lead level on intellectual development of young children. Neurotoxicol Teratol. 2000;22:805–810.

Schonfeld, D.J., Cullen, M.R., Rainey, P.M., Berg, A.T., Brown, D.R., Hogan Jr., J.C., et al. Screening for lead poisoning in an urban pediatric clinic using samples obtained by fingerstick. Pediatrics. 1994;94:174–179.

Schwartz, J. Low-level lead exposure and children's IQ: a meta-analysis and search for a threshold. Environ Res. 1994;65:42–55.

Selevan, S.G., Rice, D.C., Hogan, K.A., Euling, S.Y., Pfahles-Hutchens, A., Bethel, J. Blood lead concentration and delayed puberty in girls. N Engl J Med. 2003;348:1527–1536.

Shannon, M., Graef, J.W. Lead intoxication in children with pervasive developmental disorders. J Toxicol Clin Toxicol. 1996;34:177–181.

Shen, X.M., Guo, D., Xu, J.D., Wang, M.X., Tao, S.D., Zhou, J.D., et al. The adverse effect of marginally higher lead level on intelligence development of children: A Shanghai Study. Indian J Pediatr. 1992;59:233–238.

Silva, P.A., Hughes, P., Williams, S., Faed, J.M. Blood lead, intelligence, reading attainment, and behavior in eleven year old children in Dunedin, New Zealand. J Child Psych Psychiatry. 1988;29:43–52.

Silventoinen, K. Determinants of variation in adult body height. J Biosoc Sci. 2003; 35:263–285.

Smith, D.R., Fleegal, A.R. The public health implications of humans' natural levels of lead. Am J Public Health. 1992;82:1565–1566.

Smith, M.A. The effects of low-level lead exposure in children. In: Smith, M.A., Grant, L.D. Sors, A.I, eds. Lead Exposure and Child Development, An International Assessment. Lancaster, UK: Kluwer Publishers, 1989:1–45.

Sovcikova, E., Ursinyova, M., Wsolova, L., Rao, V., Lustik, M. Effects on the mental and motor abilities of children exposed to low levels of lead in Bratislava. Toxic Subst Mechanisms. 1997;16:221–236.

Stiles, K.M., Bellinger, D.C. Neuropsychological correlates of low-level lead exposure in school-age children: A prospective study. Neurotoxicol Teratol. 1993;15:27–35.

Stone, B.M., Reynolds, C.R. Can the National Health and Nutrition Examination Survey III (NHANES III) data help resolve the controversy over low blood lead levels and neuropsychological development in children? Arch Clin Neuropsychol. 2003;18:219–244.

Stromberg, U., Lundh, T., Schutz, A., Skerfving, S. Yearly measurements of blood lead in Swedish children since 1978: an update focusing on the petrol lead free period 1995–2001. Occup Environ Med. 2003;60:370–372.

Thane, C.W., Bates, C.J., Prentice, A. Menarche and nutritional status in pubescent British girls. Nutr Res. 2002;22:423–432.

Tong, S.L., Baghurst, P.A., McMichael, A.J., Sawyer, M.G., Mudge, J. Lifetime exposure to environmental lead and children's intelligence at 11–13 years: the Port Pirie cohort study. BMJ. 1996;312:1569–1575.

Tulve, N.S., Suggs, J.C., McCurdy, T., Cohen Hubal, E.A., Moya, J. Frequency of mouthing behavior in young children. J Expo Anal Environ Epidemiol. 2002;12:259–264.

U.S. Environmental Protection Agency. Air Quality Criteria for Lead. Research Triangle Park, NC. U.S. Environmental Protection Agency, Office of Research and Development, 1986. EPA Publication No. 600/8-83-028F. Accessed February 23, 2004.

U.S. Environmental Protection Agency. Risk Analysis to Support Standards for Lead in Paint, Dust, and Soil: Supplemental Report. Washington, D.C.: U.S. Environmental Protection Agency. 2000. EPA 747-R-00-004. Available at http://www.epa.gov/ opptintr/lead/403risksupp.htm. Accessed February 23, 2004.

U.S. Department of Health, Education and Welfare. Smoking and Health—Report of the Advisory Committee to the Surgeon General of the Public Health Service. Washington, D.C.: U.S. Department of Health, Education and Welfare, Public Health Service, 1964

U.S. Department of Housing and Urban Development. Economic Analysis of the Final Rule on Lead-Based Paint: Requirements for Notification, Evaluation, and Reduction of Lead-Based Paint Hazards in Federally-Owned Residential Property and Housing Receiving Federal Assistance. Washington, D.C.: U.S. Department of Housing and Urban Development, 1999. Available at http://www.hud.gov/offices/lead/leadsaferule/ completeRIA1012.pdf. Accessed February 23, 2004.

van Coeverden, S.C., Netelenbos, J.C., de Ridder, C.M., Roos, J.C., Popp-Snijders, C., Delemarre-van de Waal, H.A. Bone metabolism markers and bone mass in healthy pubertal boys and girls. Clin Endocrinol. 2002;57:107–116.

Walkowiak, J., Altmann, L., Kramer, U., Sveinsson, K., Turfeld, M., Weishoff-Houben, M., et al. Cognitive and sensorimotor functions in 6-year-old children in relation to lead and mercury levels: adjustment for intelligence and contrast sensitivity in computerized testing. Neurotoxicol Teratol. 1998;20:511–521.

Wasserman, G.A., Graziano, J.H., Factor-Litvak, P., Popovac, D., Morina, N., Musabegovic, A., et al. Independent effects of lead exposure and iron deficiency anemia on developmental outcome at age 2 years. J Pediatr. 1999;121:695–703.

Wasserman, G.A., Graziano, J.H., Factor-Litvak, P., Popovac, D., Morina, N., Musabegovic, A., et al. Consequences of lead exposure and iron supplementation on childhood development at age 4 years. Neurotoxicol Teratol. 1994;16:233–240.

Wasserman, G.A., Liu, X., Lolacono, N.J., Factor-Litvak, P., Kline, J.K., Popovac, D., et al. Lead exposure and intelligence in 7-year-old children: the Yugoslavia Prospective Study. Environ Health Perspect. 1997;105:956–962.

Willers, S., Schutz, A., Attewell, R., Skerfving, S. Relation between lead and calcium in blood and the involuntary smoking of children. Scand J Work Environ Health. 1988;14:385-389.

Winneke, G., Altmann, L., Kramer, U., Turfeld, M., Behler, R., Gutsmuths, F. J., et al. Neurobehavioral and neurophysiological observations in six year old children with low lead levels in East and West Germany. Neurotoxicology. 1994;15:705-713.

Winneke, G., Brockhaus, A., Ewers, U., Kramer, U., Neuf, M. Results from the European Multicenter Study on lead neurotoxicity in children: implications for risk assessment. Neurotoxicol Teratol. 1990;12:553–559.

Wolf, A.W., Jimenez, E., Lozoff, B. No evidence of developmental ill effects of low-level lead exposure in a developing country. J Dev Behav Pediatr. 1994;15:224–231.

World Health Organization, International Programme on Chemical Safety. Environmental Health Criteria 165. Inorganic Lead. Geneva: World Health Organization. 1995.

Wu, T., Buck, G.M., Mendola, P. Blood lead levels and sexual maturation in US girls: the Third National Health and Nutritional Examination Survey, 1988–1994. Environ Health Perspect. 2003;111:1–28.

Yule, W., Lansdown, R., Millar, I.B., Urbanowicz, M.A. The relationship between blood lead concentrations, intelligence and attainment in a school population: a pilot study. Dev Med Child Neurol. 1981;23:567–576.

Zhao, Q., Slavkovich, V., Zheng, W. Lead exposure promotes translocation of protein kinase C activites in rat choroid plexus in vitro, but not in vivo. Toxicol Appl Pharmacol. 1998;149:99–106.

Table R-1. Numbered references used in Tables 2 through 6.

Reference Number	First Author	Publication Date	Journal Title
1	Altmann, L.	1997	Assessment of neurophysiologic and neurobehavioral effects of environmental pollutants in 5- and 6-year-old children
2	Altmann, L.	1998	Visual functions in 6-year old children in relation to lead and mercury levels
3	Alvarez Leite, E. M.	2002	Urinary homovanillic acid and serum prolactin levels in children with low environmental exposure to lead
4	Baghurst, P. A.	1992	Environmental exposure to lead and children's intelligence at the age of seven years. The Port Pirie Cohort Study
5	Ballew, C.	1999	Blood lead concentration and children's anthropometric dimensions in the third National Health and Nutrition Examination Survey (NHANES III), 1988-1994
6	Bellinger, D. C.	1991	Low-level lead exposure and children's cognitive function in the preschool years
7	Bellinger, D. C.	1992	Low-level lead exposure, intelligence and academic achievementµ a long-term follow-up study
8	Bergomi, M.	1989	Relationship between lead exposure indicators and neuropsychological performance in children
9	Bernard, A.	1995	Renal effects in children living in the vicinity of a lead smelter
10	Calderon, J.	2001	Exposure to arsenic and lead and neuropsychological development in Mexican children
11	Canfield, R. L.	2003	Intellectual Impairment in Children with Blood Lead Concentrations below 10 µg per deciliter
12	Cooney, G. H.	1989	Low-level exposures to leadµ the Sydney lead study
13	Dietrich, K. N.	1993	The developmental consequences of low to moderate prenatal and postnatal lead exposure: intellectual attainment in the Cincinnati Lead Study Cohort following school entry

Table R-1. Numbered references used in Tables 2 through 6. *(Continued)*

Reference Number	First Author	Publication Date	Journal Title
14	Ernhart, C. B.	1987	Low level lead exposure in the prenatal and early preschool periods: early preschool development
15	Ernhart, C. B.	1988	Low level lead exposure and intelligence in the preschool years
16	Ernhart, C. B.	1989	Low level lead exposure in the prenatal and early preschool periods: intelligence prior to school entry
17	Factor-Litvak, P.	1996	Blood lead and blood pressure in young children
18	Gemmel, A.	2002	Blood lead level and dental caries in school-age children
19	Hammond, P. B.	1985	Dose-effect and dose-response relationships of blood lead to erthryocytic protoporphyrin in young children
20	Hatzakis, A.	1987	Psychometric intelligence and attentional performance deficits in lead-exposed children
21	Kafourou, A.	1997	Effects of lead on the somatic growth of children
22	Lanphear, B. P.	2000	Cognitive deficits associated with blood lead concentrations
23	McMichael, A. J.	1988	Port Pirie Cohort Study: environmental exposure to lead and children's abilities at the age of four years
24	Mendelsohn, A. L.	1999	Low-level lead exposure and cognitive development in early childhood
25	Moss, M. E.	1999	Association of dental caries and blood lead levels
26	Munoz, H.	1993	Blood Lead Level and Neurobehavioral Development among Children Living in Mexico City
27	Rabinowitz, M. B.	1986	Occurrence of elevated protoporphyrin levels in relation to lead burden in infants
28	Rahman, A.	2002	Lead-associated deficits in stature, mental ability and behaviour in children in Karachi

Table R-1. Numbered references used in Tables 2 through 6. *(Continued)*

Reference Number	First Author	Publication Date	Journal Title
29	Roels, H. A.	1987	Evaluation of dose-effect and dose-response relationships for lead exposure in different Belgian population groups (fetus, child, adult men and women)
30	Rothenberg, S. J.	1999	Pre- and postnatal lead effect on head circumference: a case for critical periods
31	Schnaas, L.	2000	Temporal pattern in the effect of postnatal blood lead level on intellectual development of young children
32	Selevan, S. G.	2003	Blood lead concentration and delayed puberty in girls
33	Silva, P. A.	1988	Blood lead, intelligence, reading attainment, and behavior in eleven year old children in Dunedin, New Zealand
34	Stiles, K. M.	1993	Neuropsychological correlates of low-level lead exposure in school-age children: A prospective study
35	Tong, S. L.	1996	Lifetime exposure to environmental lead and children's intelligence at 11-13 years: the Port Pirie cohort study
36	Walkowiak, J.	1998	Cognitive and sensorimotor functions in 6-year-old children in relation to lead and mercury levels: adjustment for intelligence and contrast sensitivity in computerized testing
37	Wasserman, G. A.	1994	Consequences of lead exposure and iron supplementation on childhood development at age 4 years
38	Wasserman, G. A.	1997	Lead exposure and intelligence in 7-year-old children: the Yugoslavia Prospective Study
39	Winneke, G.	1990	Results from the European Multicenter Study on lead neurotoxicity in children: implications for risk assessment
40	Winneke, G.	1994	Neurobehavioral and neurophysiological observations in six year old children with low lead levels in East and West Germany

Table R-1. Numbered references used in Tables 2 through 6. *(Continued)*

Reference Number	First Author	Publication Date	Journal Title
41	Wolf, A. W.	1994	No Evidence of Developmental III Effects of Low-Level Lead Exposure in a Developing Country
42	Wu, T.	2003	Blood lead levels and sexual maturation in US girls: the Third National Health and Nutritional Examination Survey, 1988-1994